International Telecommunication Standards Organizations

For a complete list of the *Artech House Telecommunication Library,*
turn to the back of this book . . .

International Telecommunication Standards Organizations

Andrew Macpherson

Artech House
Boston • London

Library of Congress Cataloging-in-Publication Data

Macpherson, Andrew.
  International telecommunication standards organizations / Andrew
Macpherson
      p.   cm.
  Includes bibliographical references.
  ISBN 0-89006-365-6
   1. Telecommunication--Standards--Societies, etc.
   2. Telecommunication--Standards--International cooperation.
  I. Title.
HE7603.M33   1990   90-41620
384'.0218--dc20   CIP

British Library Cataloguing in Publication Data

Macpherson, Andrew
  International telecommunication standards organizations.
  1. Telecommunication equipment. Standards
  I. Title
  621.38

  ISBN 0-89006-365-6

© 1990 Andrew Macpherson

Published by Artech House, Inc.
685 Canton Street
Norwood, MA 02062

International Standard Book Number: 0-89006-365-6
Library of Congress Catalog Card Number: 90-41620

10   9   8   7   6   5   4   3   2

621.380212
MACP

SUMMARY OF CONTENTS

TABLE OF CONTENTS

PREFACE

The subject of this book is literally described by its title, *International Telecommunication Organizations and Standards.* The book is directed toward the structure of the different organizations that are players in telecommunication standards, emphasizing functional standards and their importance in securing software and hardware sales in telecommunication. It focuses on international organizations and the main North American, Asian-Pacific, and European standards bodies. This book is designed to be used by a telecommunication engineer as an index to appropriate organizations, a communication manager as a means of obtaining a perspective on the various telecommunication standards players, and lawyers seeking to obtain background information for more specific advice to clients on strategies and agreements.

As standards approval has become more sophisticated and users accept the need for functional standards and standards profiles, there has emerged a link between standards and regulation as well as standards and manufacture, said link becoming more apparent as the reader uses this book. It also focuses on current issues in standards with reference to recent meetings of standards bodies.

ACKNOWLEDGMENTS

Many people from organizations around the world, particularly those included in this book, have contributed to its writing, offering suggestions for inclusion of material or checking it for accuracy. This book would not have been possible without their valuable contributions. While a detailed recitation of all involved would be too lengthy and might result in noteworthy omissions, some mention should be made of the following persons who assisted with the text or offered support: Anke Varcin, ISO; Richard E. Butler, former Secretary-General, ITU; R. Fontaine, ITU; John Francis, ITU; A. Noll, ITU; T. Irmer, CCITT; Dr. H. Ungerer, Commission of the European Communities; Mr. Wally Rothwell, ATUG; Fran Newberry and Marguerite Rona, formerly of ATUG.

In Australia, we thank Mr. Peter Walsh and Mr. R. Lions at Standards Australia; Mr.James L. Park and Dr. John Ellishaw, Telecom Australia; Mr. Bryan Bennett, Telecom Australia; Dr. Bob Horton, AUSTEL; Mr. Peter Davidson and Peter Lansdown, OTC; and officers of the Department of Transport and Communications.

In Europe, we thank P. L. Galli, CEPT; P. M. Webster, ECTEL; Rein Wolfs, RARE; Edith Hoffmann, ZVEI; Karin Garside, SPAG; Maca Jamin, OSITOP; Jerry Jayasoriya, IIC; Frede Ask, ETSI; Ruth Lawson, IFIP; John Ely, British Telecom; René Kinsoen, ECTUA; J. Rega, EWOS; George McKenderick, INTUG; D. Hekimi, ECMA; R. Desmons, IEC; and Rick Edmonson, EBEA.

In Asia, thanks go to Richard J. Barber, Pacific Telecommunications Council; T. Morikawa, Japanese Standards Association; Tadao Tamura, ITOFA; Akihide Kasukawa, Japanese Ministry of International Trade and Industry; Yuji Iwasaki, CII; E. Kinoshita, IOW; T. Watanabe, POSI.

From the United States, we thank John Mulvenna, NIST; Patti Rusher, EIA; Lori Abalta, World Federation of MAP/TOP Users' Groups; Shirley Radack, National Computer Systems Laboratory; Tim Boland, NIST; Mr. A. Lai, ESCA; Mr. Earl S. Barbely, Bureau of International Communications and Information Policy; William B. Pomeroy, International Communications Association; Trevor Housley, International Council for Computer Communication; Martha Prinsen and Mr. John Woods, IEEE Standards Office.

Much patience and encouragement has been shown by Deirdre Macpherson and Artech House, Inc. The author has appreciated talented and professional help from those at Artech House including Mark Walsh, Acquisitions Editor, Dennis Ricci, Executive Editor, and Julie Burnham, Acquisitions Assistant. Some research assistance has been given by law students at the University of Sidney Law School and particularly by Sarah Kearney. Long hours of secretarial assistance have been given by Sandra Dick-Smith, Frances Parkinson, and Julie Willard.

Andrew Macpherson

THE AUTHOR

Andrew Macpherson is a Sydney lawyer practising with Macpherson Greenleaf and Associates, a commercial law firm with specialties in technology and telecommunication law. He is a former director of the Australian Telecommunications Users Group and also has had industrial experience in database consulting and telecommunication.

CHAPTER 1

INTRODUCTION

1.1 THE TELECOMMUNICATION MIX

Much has been written about the convergence of computing and telecommunication, the former being largely unregulated and the latter firmly controlled by national post and telephone administrations.

These extreme ends of the spectrum have been moving to the center, the post and telephone administrations becoming less regulated and experiencing competition, and computing becoming more controlled where it converges with telecommunication, for example, in the area of value-added services.

Standards, too, have become more important. Their principal function is compatibility, reduction in proliferation, and control of market and trade forces. The direction is more toward the consumer and centralized network standards. Standards now play a larger role in quality control and respond to a greater extent to the needs of users and market conditions. The procedure for adopting standards must, in turn, respond to these new pressures and become more streamlined and easily adapted to users' needs.

A third factor is the need for the various professions and disciplines to see telecommunication in some perspective. Academics must mix with practitioners, radio with line, and voice with data. Overlapping areas of interest, from terminology to design, must be recognized by the different organizations and collaboration must be directly pursued.

The usefulness of separating radio and line telecommunication policy must be examined. Line telecommunication originally was the province of the post and telephone administrations because private and national boundaries were to be crossed. Radio had the same need, but, because of the relative difficulty of monitoring transmission, was far less constrained. Radio has been more of a people's medium with ham, shipping, and commercial operators all having their say. Now, as competitors and users have more influence on line communication, the interest in radio and line will become more similar. With the merging and intermingling of technologies, radio and line telecommunication must be viewed as part of a composite whole.

The link between domestic and foreign policy in telecommunication must be carefully maintained by all nations. Sovereignty questions, access to spectrum, defense, and economic development are telecommunication issues that continue with trade in information itself and many other commodities to form a delicate diplomacy mix.

Finally, there is a need to present the range of standards from both a user's and researcher's viewpoint in all its myriad aspects. In this way, a user or researcher can easily chart his or her way among the various standards bodies, study groups, or working parties to find the organization addressing his or her particular area of interest in telecommunication.

1.2 INTRODUCTION TO STANDARDS*

Government Involvement in Standards

Through involvement with the International Organization for Standardization (ISO), some 91 member countries work toward the international harmonization and development of standards. Collectively, these countries have produced over 300,000 national standards. Similarly, some 41 countries participate in the work of the International Electrotechnical Commission (IEC), which is responsible for international standardization in the electrical and electronics fields.

While these countries have similar aims in international standardization, domestically they have adopted a wide variety of standards-setting systems and procedures. Systems range from the decentralized framework that exists in the United States, where over 400 organizations produce standards of national significance, to the more formal structures of many European countries, where single national standards bodies operate. Also, there are countries such as Canada, where several standards-writing organizations administer codes under the aegis of a central coordinating statutory corporation, the Standards Council of Canada.

The status of national standards organizations in any particular country depends on the level of government mandate or voluntary recognition. In most countries, the national standards bodies have obtained a charter or some other formal document from the national government giving standards-writing authority. One notable exception is the United States, where the

---

* See Report of the Committee of Review of Standards, Accreditation, and Quality Control and Assurance, chaired by Mr. Kevin Foley and prepared for Senator John Button of the Ministry of Industry, Technology and Commerce, Parliament of the Commonwealth of Australia, June 1987.

American National Standards Institute (ANSI), the organization formally linked with ISO and IEC, has been unable to obtain such authority, possibly due to the large number of existing standards-writing bodies, both in the private sector and branches of the federal government. ANSI, composed of corporations, standards developers, government bodies, and individuals, regards itself as the national standards body on the basis that it provides and administers the only recognized system in the United States for establishing American National Standards. However, major organizations, such as the American Society for Testing and Materials (ASTM), the Society of Automotive Engineers (SAE), the Institute of Electrical and Electronics Engineers (IEEE), and the American Petroleum Institute (API) produce nationally and internationally recognised standards. ASTM accounts for approximately 7200 of a total of 32,000 US voluntary standards.

The Federal Republic of Germany (FRG) has a contract with the German Standards Institution (Deutsches Institut für Normung e.V.) in which the government formally recognises DIN as the competent standards organization for the FRG and applies DIN standards in contract documentation.

Similarly, the United Kingdom saw the need for a more formal and detailed understanding between the government and the British Standards Institute (BSI), and a Memorandum of Understanding was signed in 1982.

The level of funding provided by governments to their national standards organizations varies significantly. In OECD countries, the average rate of government funding of comparable standards bodies is around 30 per cent. However, in Japan, the national standards body, the Japanese Industrial Standards Committee (JISC), relies on 100 per cent government funding. At the other end of the scale, ANSI in the United States and the Association Suisse de Normalisation (SNV) in Switzerland receive no government funding, relying almost entirely on subscriptions and sales of publications for their revenue.

Compliance with Standards

There are varying degrees of compulsion attached to standards in different countries. In France, in the execution of central and local government procurement contracts, and any government-subsidized or publicly funded projects, compliance with standards may be optional, although this is becoming less so. If a standard is expressly mentioned in a contract, application is usually legally binding. In addition, ministerial approval may include a decision to make a standard mandatory. Under the Napoleonic Code, the designer and manufacturer of a product are legally liable for its satisfactory performance for a considerable time after its sale (e.g., ten years in the case of buildings). To insure against this liability is customary, and as a result insurance companies require certification of compliance with standards or testing by an independent approval body as a condition of insurance.

In the Federal Republic of Germany, DIN standards are noncompulsory recommendations, although many customers prefer products that meet these specifications, and government authorities often cite standards in administrative instructions. In practice, through testing and approval processes, the German market requires that the manufactured products comply with DIN standards, especially where safety is an issue.

In Japan, the Japanese Industrial Standards (JIS) mark is a voluntary system to certify conformity of a product's performance and quality to a specified standard. There is no prohibition of unmarked products and no compulsion for manufacturers to belong to the JIS marking scheme. However, application of the JIS mark is strictly regulated, and penalties are applied to unjustified use. The Japanese Ministry of International Trade and Industry (MITI) may designate products as falling under the JIS marking system. Manufacturers who wish to use the JIS mark must then apply to MITI for a license to do so. Before granting the license, MITI will examine their manufacturing and quality control systems and require third-party testing of the product by an accredited laboratory. In effect, the government controls the use of the JIS mark, but the standards are written by JIS and adopted by MITI on the recommendation of JIS. Approval of products and assessment of a manufacturer's quality capability are done by independent testing houses. The accreditation of testing laboratories is granted directly by MITI.

In New Zealand, all standards are voluntary at the time that they are declared by the Standards Association of New Zealand (SANZ), but a number have become mandatory through incorporation into government legislation or local authority bylaws.

About 300 of the 8500 standards in the United Kingdom receive some kind of official backing. In some UK regulations, compliance with a British Standard product specification is made a requirement of the regulations, and in others, British Standard measurement and testing methods are cited in regulations, which themselves stipulate the performance to be attained. Despite mechanisms for referring to standards in legislation, the United Kingdom, unlike many other countries, has traditionally made little use of standards and certification in primary legislation or in regulations. As a result, British industry is not generally accustomed to working within such constraints.

US manufacturers expect to demonstrate compliance with a variety of national and local requirements, especially those concerned with safety. On a nationwide basis, Underwriters Laboratories, Inc. (UL), a nonprofit, independent organization, operates certification schemes for its own standards, about half of which have been adopted by ANSI. The Factory Mutual Research Corporation (FM) also operates approval schemes for equipment, materials, and services in the electrotechnical and construction fields, which principally bear upon safety from fire. FM and other similar organizations apply their own standards or those produced by other organizations. Both FM and UL are linked with the insurance industry, whose policies frequently require that approved products be used in buildings and equipment.

The New Rules for International Standardization

In the past, national standards have been developed and presented to the national body for recognition as national standards. These, in turn, have then been used as draft proposals for international standards in ISO and IEC. In those cases in which the United States has dominated the marketplace, the US standard has become the basis for the international standards. When some other entity dominated the marketplace, the standard developed by that group or country was more likely to become the basis for the international standard.

The rules for international standardization have now changed and are no longer the basis of market dominance alone. Technological dominance may even replace market dominance as the key factor leading to the acceptance of a proposed standard. In 1992, for the the first time there will be two major markets in the world — Europe and America — and maybe three in the future if the Pacific Rim or Asian Belt countries develop. Market dominance by a single national or economic interest is no longer possible. An effective standards program will require international involvement from the beginning, as being international first and national second is cost effective and economically essential.

Standardization has moved from an evolutionary process of common use and acceptance of long-established practice (a bottom-up process) to one of standardization of essential elements, interfaces, compatibility requirements, and configuration controls before products are ever completely developed (a top-down process).

Such groups as the IEEE are well placed to handle the internationalization of standards because the IEEE's committees and standards-making process are transnational.

The ISO Industry View

Lawrence D. Eicher, Director General of ISO, has expressed the view that the standardization process has moved from the postmanufacturing stage to the research and development side of new products. This view is set out in *The IEEE Standards Bearer,* June 1989, Vol. 3, No. 3. Eicher indicates that science, by its nature, makes progress geometrically. A given level of worldwide scientific effort can be expected to produce progress proportionally to the state of current knowledge — the calculus of discovery produces exponential benefits.

Prior to World War II, engineering progress tended to be linear, deriving more from experience than from theory. Since the late 1950s, most fields of engineering have become scientific. Designs, processes, and both product and system functions are now modeled, predicted, specified, and controlled. The innovation process must now be seen as a series of concurrent interactive processes with heavy dependence on basic science and scientific engineering at every step.

This progressive development has meant that there have been a number of fundamental changes to industry and the emergence of a new industrial economy.

The main features of this economy are
- the emergence of controlled transnational production systems capable of rapid and concerted evolution,
- a shift of key elements of production control from design specifications and mechanical parts of machines to performance requirements for processes and interfaces,
- a release of opportunities for innovation (to enhance end-user satisfaction) from the constraints of scale economies in production, and
- the emergence of complementary assets (such as consumables, software, education, or support services) as concurrent necessities to develop new global markets for industrial products.

Research and Development as Part of Standardization

Total international trade in industrial products is growing much faster than the sum of Gross Domestic Products (GDP), illustrating the reality of the emerging global industrial marketplace. Obviously, global standardization grows in importance relative to domestic standardization because only global solutions are satisfying to geographically dispersed, vertically integrated industries. International harmonization of standardization is the only way to lessen the temptation for arbitrariness in domestic or regional standards, which may reflect protectionist and political pressures.

The ISO-IEC system, like many national systems, is based almost totally on the consensus concept, involving users and producers seeking solutions in their common interest, but an emerging technology has no users yet! Nevertheless, standardization is already underway, although the standardizers usually do not think of themselves as such. In the very early stages of development of a new technology, applications can be imagined, but functional prototypes do not exist. Most of the items in the ISO Long Range Planning Survey fall into this category. Hence, the need for standardization is restricted to terminology and derives mainly from the basic sciences.

When functionality has been demonstrated and market entry is anticipated, problems of technological choice arise, and needs for an early phase of international agreements begin to emerge. The focus is on

- Health, safety, and environmental concerns of the public need to be addressed. Standards can only predict possible harm not yet experienced, and knowledge is often faulty. Standardization efforts should focus on accumulating databases of quantitative information for use in control regimes to be defined later;
- For many new consumer products requiring compatible media, software, or usable goods, international agreements are critical to market development (e.g., the laser dish, but not razor blades);
- International agreements are also key to the utility of new products with interface specifications to national and regional technology infrastructures that need to be harmonized (e.g., high definition television, HDTV);
- Agreements on interface reference models (e.g., the ISO Open System Interconnection standard, OSI) are exceptionally beneficial at this early stage with products for which market demands for interoperability can be foreseen (information technology in general).

As new markets begin to mature, the marketplace itself begins to define such beneficial standardization opportunities as variety, simplification, and improved utility. Product standardization becomes more important, but intelligent product standardization at this stage is difficult.

Full-fledged consensus procedures are most suitable for working on issues at the product standardization stage, when international agreements may affect industrial infrastructure and national interests. These procedures are often too cumbersome and time-consuming to be useful during premarket stages of development. Consequently, industry experts participating in consensus standards work are normally not the same people who work on future product designs.

Nevertheless, for the R&D community, early focusing on standards questions through sharing of information, collaborative research, open publication, and direct professional interaction clearly would be extremely beneficial, independently of the need for consensus to emerge at a later date.

Much could be achieved if ISO and IEC, together with their network of global standardization partners, viewed their role as including promotion of research knowledge and its diffusion, with the long-range view of facilitating the achievement of consensus agreement when the time is right.

1.3   USING THIS BOOK

Philosophy of the Book

This book has been designed for engineers, communication managers, and other professionals who need to be familiar with organizations and published materials on telecommunication standards. The book should be particularly useful where telecommunication involves cross-fertilization of ideas in different disciplines. Engineers in one discipline can learn about related disciplines and persons with nontechnical backgrounds can ascertain the best method of accessing information on telecommunication for achieving their objectives. This book particularly recognizes that voice telecommunication and data communication have been considered separate disciplines as have the physical line and air media. Radio is mainly voice communication with some video. Information processing, however, often requires a reasonable understanding of both media and further brings users of applications into contact with the media themselves. There is very little use of telecommunication that is entirely transparent to the user.

The book concentrates on "what is," rather than "what might be," and seeks to bridge the gap between a sociological textbook and a technical reference manual. In particular, it is designed to direct the reader to telecommunication organizations or standards that directly and suitably address his or her problem or application.

Use as a Reference Tool

The format is shown in the short summary of contents. The book comprises a mixture of information on telecommunication standards organizations and the making of standards. Each chapter is divided into numbered sections. Where necessary, the sections are further divided into subsections.

The book has been designed to be read as a whole or used by reading selected sections or chapters. As each chapter stands alone, there is some overlap with others. The chapters have numbered sections that can also be read in isolation to assist the reader's purpose.

The short summary and the expanded table of contents have been designed to give the reader direct access to telecommunication organizations and individual features of their telecommunication standards. Using the short summary and then the expanded table of contents should be the quickest way to find relevant information. The glossary will also be useful.

The book's ultimate objective is to give direction on how standards in various areas may be obtained.

CHAPTER 2

THE INTERNATIONAL
TELECOMMUNICATION UNION

International Telecommunication Union (ITU)
Place des Nations
CH–1211 Genève 20
Switzerland

Telephone:	41 22 730 5111
Facsimile:	41 22 733 7256
Teletex:	228-468 15100 = UIT

2.1 THE ORIGINS OF THE ITU

The main international telecommunication body is the International Telecommunication Union (ITU), also known as the Union Internationale des Télécommunications (UIT).

Founded in Paris in 1865 as the International Telegraph Union, the International Telecommunication Union took its present name in 1932 and became a Specialized Agency of the United Nations in 1947. The ITU is governed by the International Telecommunication Constitution and Convention, which was adopted at Nice, France, in June 1989.

2.2 THE ITU AND THE UNITED NATIONS

The principal organ of the United Nations (UN) is the General Assembly, which comprises its 159 member states.

There are five other major organs of the UN, one of which is the Economic and Social Council. The members of this council number 54 and are elected by the General Assembly. The role of the Economic and Social Council is to assist the UN in fostering higher standards of living as well as social and economic progress.

To achieve these aims, the council can call upon the aid of specialized agencies outside the UN. The provisions in Article 57 and Article 63 of the UN Charter, which refer to specialized agencies, are extracted as follows:

The various specialized agencies, established by intergovernmental agreement and having wide international responsibilities, as defined in their basic instruments, in economic, social, cultural, educational, health and related fields, shall be brought into relationship with the United Nations in accordance with the provisions of Article 63.

Such agencies thus brought into relationship with the United Nations are hereinafter referred to as "specialized agencies." ...

The Economic and Social Council may enter into agreements with any of the agencies referred to in Article 57, defining the terms on which the agency concerned shall be brought into relationship with the United Nations. Such agreements shall be subject to approval by the General Assembly.

It may co-ordinate the activities of the specialized agencies through consultation with and recommendations to such agencies and through recommendations to the General Assembly and to the Members of the United Nations.

The International Telecommunication Union is one of the specialized agencies of the United Nations.

At a Plenipotentiary Conference held in the United States of America in 1947, the ITU adjusted its organizational structure and entered into an agreement with the UN whereby the ITU was recognized as the specialized agency for telecommunication. A new convention was adopted in 1947 to effect those changes. Since then, the ITU Convention has been revised and updated on several occasions, being divided into a Constitution and a Convention in 1989.

2.3 THE WORK OF THE ITU

The ITU is responsible for the regulation and planning of worldwide telecommunication, the establishment of operating standards for equipment and systems, the coordination and dissemination of information required for the planning and operation of telecommunication services, and the promotion of and contribution to the development of telecommunication and related services. The ITU Constitution states that the purposes of the ITU are to

- maintain and to extend international cooperation for the improvement and rational use of telecommunication of all kinds;
- promote and to offer technical assistance to developing countries in the field of telecommunication;
- promote the development of technical facilities and their most efficient operation with a view to improving the efficiency of telecommunication services, increasing their usefulness, and making them, so far as possible, generally available to the public;
- promote the use of telecommunication services with the objective of facilitating peaceful relations; and
- harmonize the actions of members in the attainment of those common ends.

In particular the ITU is to

- effect allocation of the radio frequency spectrum with a view to avoiding harmful interference between radio stations of different countries;
- allot radio frequencies and to register frequency assignments and associated geostationary-satellite orbital positions;
- coordinate efforts to eliminate harmful radio station interference;
- improve the use of the radio frequency (RF) spectrum for radiocommunication services;
- improve the use of the geostationary satellite orbit (GSO) for radiocommunication services;
- facilitate the worldwide standardization of telecommunication, with a satisfactory quality of service;
- foster international cooperation in the delivery of technical assistance to developing countries;
- coordinate efforts to harmonize the development of telecommunication facilities;
- foster collaboration among its members with a view to establishment of rates at levels as low as possible, consistent with an efficient service;
- promote the adoption of measures for ensuring the safety of life through the cooperation of telecommunication services;
- undertake studies, make regulations, adopt resolutions, formulate

recommendations and opinions, and collect and publish information concerning telecommunication matters; and

- promote, with international financial organizations, credit for the development of social telecommunication projects in isolated areas.

The ITU is working to achieve international agreement on equitable access to the radio frequency spectrum and the geostationary satellite orbit as well as global interconnection and interoperation of telecommunication.

The resolution of disputes is achieved in three main ways, namely by

- international conferences and meetings,
- dissemination of information including world exhibitions, and
- technical cooperation.

The Regulatory Function

The ITU incorporates regulatory, standardization, and development functions. The regulatory function includes establishing technical procedures for the coordination, notification, and recording of frequency and orbital assignments to prevent or eliminate harmful interference between radio stations of different countries and to make more efficient use of the radio frequency spectrum and the geostationary satellite orbit. The regulatory function also involves the establishment of procedures for the interconnection and interoperation of international networks and services.

The Standardization Function

The standardization function was initially required for telephony between countries, as cited in the International Telecommunication Regulations. This function has expanded to cover numerous aspects of a range of long-distance telecommunication services.

The Development Function

The development function is concerned with promoting the development of technical facilities and the improvement of telecommunication networks, particularly in the Third World. Developing countries, propelled by their growing reliance on telecommunication to foster economic growth and political unity, argue vociferously for large increases in technical assistance resources.

Table 2.1. Contents of the Constitution and Convention of the International Telecommunication Union

Table of Contents

Constitution of the International Telecommunication Union

Convention of the International Telecommunication Union

To satisfy more fully the requirements of different countries, the world is divided into three ITU regions. In broad terms, Region 1 covers Europe (including the USSR) and Africa, Region 2 comprises the Americas, and Region 3 encompasses Asia and Oceania. Regional Administrative Conferences may be held to reach agreement on regional issues. The agenda of a Regional Administrative Conference is restricted to the region concerned and must be in conformity with the Administrative Regulations.

World Administrative Radio Conferences (WARCs) are held as decided by the Plenipotentiary Conference, and have the authority to consider and to approve changes to the Radio Regulations.

World Administrative Telegraph and Telephone Conferences (WATTCs) have similar powers with respect to the Administrative Regulations relating to telegraphy and telephony. A WATTC held at Melbourne, Australia, in 1988 adopted new Telecommunication Regulations to replace the previous International Telegraph and Telephone Regulations. These conferences are discussed in Chapter 3.

Exhibitions

The General Secretariat, at the request of the Plenipotentiary Conference, also organizes telecommunication exhibitions, which are intended "to keep members of the Union informed of the latest advances in telecommunication techniques." The ITU has organized a world exhibition, TELECOM (telecommunication exhibition), every four years, and in 1989 held ITU-COM (electronic media exhibition). It also cosponsors with telecommunication administrations of ITU members regional exhibitions (ASIA TELECOM, AFRICA TELECOM, AMERICAS TELECOM).

2.7 INTERNATIONAL CONSULTATIVE COMMITTEES

The International Radio Consultative Committee (CCIR) and the International Telegraph and Telephone Consultative Committee (CCITT)

The two international consultative committees are separate bodies respectively dealing with technical questions in radio and in telegraph and telephone. All member countries of the ITU can participate in their work, as can certain private companies providing telecommunication technology and services.

Each CCI holds a Plenary Assembly every four years. The Plenary Assembly draws up a list of technical telecommunication subjects or "questions," the study of which leads to improvements in international radiocommunication or international telegraphy and telephony. These questions are then entrusted to a number of Study Groups, composed of experts from different countries, scientific or industrial organizations, international organizations, and regional telecommunication organizations. The Study Groups draw up recommendations, which are submitted to the next Plenary Assembly. If the Assembly adopts the recommendations, they are published. CCIR and CCITT recommendations have an important influence on telecommunication engineers and technicians, operating administrations and companies, and manufacturers and designers of equipment throughout the world.

The CCITT and the CCIR each have a director elected by the Plenipotentiary Conference, whose functions include organizing the work of the consultative committee (e.g., Study Group meetings, Plenary Assembly, technical documents). The director is assisted by a specialized secretariat consisting of about 30 or 40 engineers, editorial, and other professional staff.

The Consultative Committees are discussed in greater detail in Chapters 4 and 5.

World Plan Committees

The consultative committees organize a World Plan Committee and such Regional Plan Committees as may be jointly approved by the Plenary Assemblies of the international consultative committees. The Plan Committees develop a General Plan for the international telecommunication network to facilitate coordinated development of international telecommunication services. They refer to the CCITT and CCIR questions which are of particular interest to developing countries. There are three Regional Plan Committees for Latin America, Africa, Asia and Oceania, Europe and the Mediterranean Basin.

The CCIs have Special Autonomous Groups (GAS), set up to consider specific questions of particular interest to developing countries.

2.8 THE INTERNATIONAL FREQUENCY REGISTRATION BOARD (IFRB)

The International Frequency Registration Board (IFRB) consists of five independent radio experts from different regions of the world, elected by the Plenipotentiary Conference, who work full time at the ITU's headquarters in Geneva. These members do not represent their regions, but rather act as custodians of an international public trust, involving management of the world's radio frequency spectrum.

The IFRB's main task is to decide whether radio frequencies which the countries assign to their radio stations are in accordance with the convention and the Radio Regulations, and whether these frequency allocations may cause harmful interference to other stations. If the IFRB's finding in a particular case is favorable, the frequency is recorded in the huge Master Frequency Register kept by the IFRB, and thus obtains formal international recognition and protection. The Master Frequency Register contains over 1,000,000 assignments (1988) and an average of 1200 assignment notices (covering new assignments or modifications to existing assignments) arrive at the IFRB each week.

Among the major tasks of the IFRB are

- the provision of advice to members, particularly on frequency management and the use of the radio frequency spectrum and the geostationary satellite orbit;
- participation at the request of governments in the obligatory intergovernment coordination of the use of frequencies involving space services prior to their notification for recording in the Master Frequency Register;
- the orderly recording of the positions assigned by countries to geostationary satellites;
- technical assistance in the preparation of radio conferences; and
- the preparation of publications about its work.

The IFRB is considered further in Appendix 2B.

2.9 ADMINISTRATIVE ORGANIZATION

Administrative Council

The Administrative Council is composed of 43 ITU members elected by the Plenipotentiary Conference. The Administrative Council normally meets for about two weeks once a year at ITU headquarters in Geneva. The council supervises the administrative functions and coordinates the activities of the five permanent organs at the headquarters, and examines and approves the annual budget at formal sessions. The council acts for the Plenipotentiary Conference between its meetings. The detail of the Administrative Council's functions is set out in Article 10 of the ITU Constitution.

General Secretariat

The General Secretariat is directed by the Secretary-General and is responsible for the administrative and financial services of the ITU and for application of staff and financial regulations as approved. The Secretary-General is also responsible for operational matters, information exchange, and external relations, and acts as the legal representative of the ITU.

Panels of Experts

Panels of Experts are convened to examine particular matters of importance to the ITU. These panels report to the Administrative Council. There were, prior to the Nice Plenipotentiary Conference, two such panels active; one dividing the International Telecommunication Convention so as to divide it into a Constitution and separate Convention, and the other examining the long-term future of the International Frequency Registration Board.

Coordination Committee

The Coordination Committee comprises the Secretary-General, Deputy Secretary-General, Directors of the CCITT and CCIR, and Chairman and Vice-Chairman of the IFRB.

The Coordination Committee advises the Secretary-General and gives practical assistance on all administrative, financial, and technical cooperation matters affecting more than one permanent organ, and external relations and public information.

Telecommunications Development Bureau (BDT)

Prior to the Nice Plenipotentiary Conference the Technical Cooperation Department (TCD) of the General Secretariat administered, mainly within the framework of the United Nations Development Program (UNDP), a program through which telecommunication projects were undertaken in various countries throughout the world on the planning and operation of telecommunication systems and human resource development.

On average, the ITU carries out between 170 and 190 projects every year in about 100 developing countries, representing from $25 to $28 million. The technical cooperation projects are mainly financed from the United

Nations Development Program (85% of the total number of projects) and Funds-in-Trust (15%).

The Nice Plenipotentiary Conference established the Telecommunications Development Bureau (BDT) within the structure of the ITU, with the same status as the other permanent organs (General Secretariat, CCIs, and IFRB), for the purpose of stengthening technical cooperation. The BDT absorbs the TCD. This act marked the culmination, in practical terms, of the initiative taken at the 1982 Nairobi conference, which enshrined technical cooperation and assistance to the developing countries in the International Telecommunication Convention as one of the ITU's basic purposes.

The BDT will be responsible for performing the ITU's dual role as the United Nations specialized agency within its sphere of influence. The BDT will be free to make other financing arrangements to facilitate telecommunication development, and will benefit from stable and regular funding through the ITU's ordinary budget.

Center for Telecommunications Development

In 1987, a new dimension was added to the ITU's technical cooperation activities. The Center for Telecommunications Development, set up in 1985, started its field operations. The center is funded entirely through voluntary contributions from the public and private sectors. The center's goals are to help developing countries raise sufficient resources on a stable and continuous basis to strengthen their social and development plans. The Plan of Action includes the assistance of countries in the identification and formulation of their project proposals for funding by banking and other financial institutions. The center has not been successful in raising funds, thereby giving rise to greater support for the BDT. The center's future eventually will be reconsidered by the Administrative Council.

Networks

Through the assistance of the TCD, modern regional telecommunication networks have been or are in the process of being set up in Latin America, Asia, Africa (PANAFTEL), and the Mediterranean and Middle East (MEDARAB-TEL), and numerous national networks have been upgraded and expanded. New major projects include TELDEV (Telecommunications Development in six Arab States) and RASCOM (Regional African Satellite Communication project) now at the stage of feasibility study to define the parameters and characteristics of the system to be installed.

RASCOM is a total telecommunication network study for Africa, commissioned by the African Ministers of Planning, Transport, and Communications to ascertain the long-term telecommunication requirements and to provide efficient infrastructures for the development of the continent as a whole. RASCOM puts special emphasis on the rural areas where the majority of the continent's population resides. Furthermore, RASCOM approaches the problem of access to telecommunication services not just as traditionally perceived by the PTT authority, but also as viewed by the overall national development authority and users.

Moreover, valuable assistance has been rendered to a number of administrations, especially in Africa, by drawing up Master Plans for the long-term development of their networks. These Master Plans cover all aspects of the development plan, including technical, manpower, and training requirements.

Outer Space

Several world administrative conferences for space telecommunication have been held as well as a world administrative conference on satellite broadcasting.

2.10 PUBLISHING

General Information

The General Secretariat assembles international telecommunication data published for the benefit of telecommunication engineers and operating authorities. There are lists of radio stations and telegraph offices throughout the world, teaching aids for training institutions, statistics, maps, charts and tables, and a monthly journal.

Telecommunication Journal

The *Telecommunication Journal* is the monthly magazine of the ITU, published in English, French, and Spanish. Annual subscription by surface mail is 90 Swiss francs (CHF) yearly. Airmail delivery is an additional CHF 48 in the European zone and CHF 96 anywhere else in the world.

Subscriptions should be ordered from:

> Subscriptions
> La Presse Technique SA
> 3a, rue du Vieux-Billard

CH–1205 Genève

Switzerland

The *Telecommunication Journal* contains an editorial, details of ITU activities, major articles, CCIR news, CCITT news, particulars under the headings of Technical Cooperation, Official Announcements, Calendar of ITU Conferences and Meetings, and ITU Publications and Conferences or Meetings External to the ITU. There is often a section on ideas and achievements. A sample content page from the *Telecommunication Journal* can be seen in Appendix 2C.

2.11 SOME POLICY ISSUES

The 1982 ITU Plenipotentiary Conference held in Nairobi was a significant international instrument for the conduct of international telecommunication policy. Geopolitical, organizational, and technical issues at the ITU combine to form a broad foreign policy agenda in telecommunication. The ITU's work is a continual process of shaping and modifying foreign policies so that there evolve more definite and consistent telecommunication foreign policies in the long term.

Smaller nations worry about domestic economic dependence on telecommunication and the growth of a technological colonialism. In particular, developing nations are conscious of the economic costs of assessing technology, and their access to the electromagnetic spectrum and the geostationary satellite orbit. Countries need to manage a dynamic environment and to harness the opportunities created by advances in technology to continue the relatively high level of order and dynamic stability that has prevailed since World War II. The sources of stability are strengthened at a comparatively slow rate.

The ITU has long had preeminence in the field of international telecommunication standards-making, particularly because of its nonpartisan United Nation status and the wide spread of its membership, covering a geographic and developmental spectrum of member countries. In later times, geographic groupings, such as Europe, Oceania, and North America, because of the importance of telecommunication standards-making in manufacturing and trade involving telecommunication services, have sought to exert greater control over standards-making. This has meant a need for increasing the speed of standards-making and the procedures previously involved in both the CCITT and the CCIR have been or are being appropriately improved.

The ITU's current move to restructure both management and organization with a view to holding its preeminence and control in international telecommunication standards-making has been necessary and is likely to achieve that result.

Notwithstanding its standards-making role, the ITU has a unique position in setting telephone rates. While nations may address rates separately between each other to some extent under the Regulations, the ITU's importance as a forum for this purpose is unlikely to be challenged in the near future. The ITU continues to play a vital role in the registration and administration of radiocommunication frequency assignments.

2.12 OPERATIONAL STRUCTURE AND MEETINGS STRUCTURE

Table 2.2 (The International Telecommunication Union Administrative Structure) and Table 2.3 (International Telecommunication Union Meetings Structure), which follow show in diagrammatic form the working of the ITU.

Table 2.2. The International Telecommunication Union Operational Structure

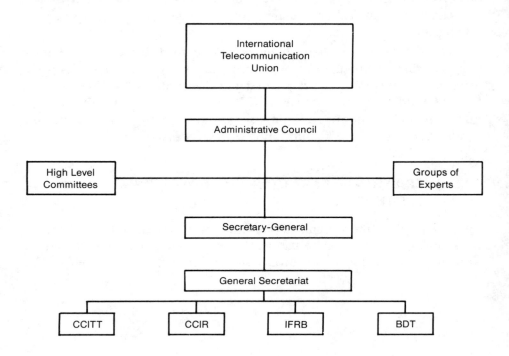

CCITT   International Telegraph and Telephone Consultative Committee
CCIR   International Radio Consultative Committee
IFRB   International Frequency Registration Board
BDT   Telecommunications Development Bureau

Table 2.3. International Telecommunication Union Meetings Structure

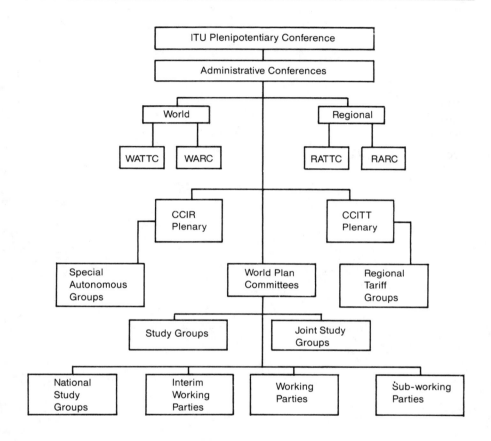

WATTC   World Administrative Telegraph and Telephone Conference
WARC   World Administrative Radio Conference
CCIR   International Radio Consultative Committee
CCITT   International Telegraph and Telephone Consultative Committee

APPENDIX 2A

TABLE OF CONTENTS
CHAPTERS AND ARTICLES
CONSTITUTION AND CONVENTION
OF THE INTERNATIONAL
TELECOMMUNICATION UNION

DECLARATIONS AND RESERVATIONS made at the end of the Plenitotentiary Conference of the International Telecommunication Union (Nice, 1989).

Optional Protocol to the Constitution of the International Telecommunication Union and to the Convention of the International Telecommunication Union on the Compulsory Settlement of Disputes.

APPENDIX 2B

THE INTERNATIONAL FREQUENCY REGISTRATION BOARD (IFRB) AND THE RADIO REGULATIONS

2B.1 THE INTERNATIONAL FREQUENCY REGISTRATION BOARD (IFRB)

Chapter IV of the Radio Regulations is titled, "Coordination, Notification and Registration of Frequencies. International Frequency Registration Board."

This chapter contains Articles 10 to 17, and Article 10 is titled, "International Frequency Registration Board." Article 10 has two sections, namely, Section I—Functions of the Board and Section II—Methods of Work of the Board.

The 16 functions of the board are encompassed in Regulations 991 to 1007 and the methods of work of the board in Regulations 1008 to 1016.

The functions of the IFRB include

- the processing of frequency assignment notices, including notices about any associated orbit or location of geostationary satellites, received from administrations for recording in the Master International Frequency Register,
- the processing of information received from administrations in application of the advanced publication, coordination and other procedures of the Radio Regulations and the Final Acts of the administrative radio conferences,
- the processing and coordination of seasonal schedules of high frequency broadcasting, the compilation for publication by the Secretary-General of frequency lists reflecting the data recorded in the Master International Frequency Register,

- the study, on a long-term basis, of the usage of the radio frequency spectrum with a view to making recommendations for its more effective use,
- the investigation at the request of one or more interested administrations of harmful interference,
- the provision of assistance to administrations in the field of radio spectrum utilization,
- the development of Technical Standards in accordance with Radio Regulations, numbers 1454 and 1582, and the Rules of Procedure for internal use by the IFRB in the exercise of its functions,
- the formulation and reference to the CCIR of all general technical questions arising from the IFRB's examination of frequency assignments,
- the technical assistance in the preparation for an organisation of radio conferences,
- the participation in an advisory capacity, upon invitation, in conferences and meetings relating to the assignment and utilization of frequencies,
- the provision of assistance to administrations in training of senior staff in spectrum management and utilization, and
- the discharge of such other functions as are specified in the Radio Regulations and in the Final Acts of administrative radio conferences.

Methods of Work of the IFRB

The IFRB consists of five members and usually meets once a week. It elects each year a Chairman and Vice-Chairman for a one-year term. Each member of the board has one vote. However, there is no voting by proxy or by correspondence. The minutes indicate whether a decision was unanimous or by a majority.

A quorum is one-half of the members of the IFRB and the board is to endeavour to reach its decisions by unanimous agreement.

The documents of the IFRB, which comprise a complete record of its official actions and minutes of its meetings, are maintained by the board and are available for public inspection at its offices.

Technical Standards

Radio Regulation 1454 is as follows:

The Technical Standards of the Board shall be based on the relevant provisions of these Regulations and the Appendixes, the decisions of administrative conferences of the Union, as appropriate, the Recommendations of the CCIR, the state of the radio art and the development of new transmission techniques, account being taken of acceptable propagation conditions which may prevail in certain regions.

Radio Regulation 1001.1 is on the same subject:

The Technical Standards and the Rules of Procedure of the IFRB are distributed to all Members of the Union and shall be open to comment from any administration. In the event of there being a disagreement which remains unresolved, the procedure to be followed is given in Resolution 35.

2B.2 RADIO REGULATIONS

The radio regulations comprise two ring-bound loose-leaf volumes. The Radio Regulations are published under the authority of the Secretary-General of the International Telecommunication Union. It is a consolidated document, which incorporates, in Volume 1, the provisions of the Radio Regulations (Geneva 1979) and Appendices 1–24 thereto, and, in Volume 2, Appendices 25–44, as well as resolutions and recommendations, as adopted by the World Administrative Radio Conferences. Volume 1 also contains the Table of Contents, a Foreword, an Analytical Table, an Analytical Index, and Notes.

This edition includes the partial revisions of 1985, 1986, and 1988 adopted respectively by the following conferences:

- World Administrative Radio Conference for the Mobile Services, Geneva, 1983 (MOB-83),
- First session of the World Administrative Radio Conference on the Use of the Geostationary-Satellite Orbit and the Planning of Services Utilizing It, Geneva, 1985 (ORB-85),
- World Administrative Radio Conference for the Planning of the HF Bands Allocated to the Broadcasting Service, Geneva, 1987 (HFBC-87).

The Table of Contents sets out the Radio Regulations, which are in two parts and divided into chapters, articles, and sections. The framework by way of chapters is as given below.

RADIO REGULATIONS

Preamble
PART A

CHAPTER	HEADING
I	Terminology.
II	Technical Characteristics of Stations.
III	Frequencies.
IV	Co-ordination, Notification Registration of Frequency. International Frequency Registration Board.
V	Measures against Interference. Tests.
VI	Administrative Provisions for Stations.
VII	Service Documents.
PART B	
VIII	Provisions Relating to Groups of Services and to Specific Services and Stations.
IX	Distress and Safety Communications.
X	Aeronautical Mobile Service and Aeronautical Mobile-Satellite Service.
XI	Maritime Mobile Service and Maritime Mobile-Satellite Service.
XII	Land Mobile Service.
XIII	Entry into Force of the Radio Regulations.
APPENDICES 1–44	
	Resolutions.
	Recommendations.

Chapter IV on provisions relating to international frequency management is divided into Article 10 and Articles 11, 12, 13, 14, 15, 15A, 16, and 17. Article 10 is on the IRFB (see Section 2B.1 of this Appendix).

In Chapter VIII, provisions relating to "Groups of Services and to Specific Services and Stations," there are Articles 27 to 36, among which Article 30 covers broadcasting service and broadcasting satellite service.

The Analytical Table in Volume 1 is in the form of an index with references to Annexes, Appendices, Articles, Resolutions, Recommendations, and Sections. The Analytical Index of Resolutions and Recommendations is a simplified method of checking the substance of the Resolution or Recommendation by an index which is related to the number of the Resolution or Recommendation.

The Notes by the General-Secretary include:

- Call sign formation possibilities,
- Provisions of the Radio Regulations containing references to CCIR Recommendations, and
- Flow charts extracted from the IFRB Handbook on Radio Regulatory Procedures.

There are 17 flow charts. The Recommendations then follow in full.

An example of the appendices follows:

APPENDIX 1 — Basic characteristics to be furnished for notification under the radio regulations including information, form of notice and general instructions.

APPENDIX 2 — Submission of HF Broadcasting requirements to the IFRB.

APPENDIX 3 — Notices relating to Space Radio Communications and Radio Astronomy Stations.

The Recommendations and Resolutions are arranged in numerical order and numbered along the lines of the groupings and numbering systems given in the sample page reproduced here.

Note by the General Secretariat

The Resolutions are arranged in order and numbered along the lines of the grouping and numbering system below. As some Resolutions in one group have direct relationship to Resolutions in other groups, this has been reflected, as far as possible, to facilitate consultation.

In this respect, see also the Analytical Index.

(Rev. 1986)

APPENDIX 2C

SAMPLE CONTENTS PAGE FROM THE TELECOMMUNICATION JOURNAL

Contents

VOLUME 56 – No. X
OCTOBER 1989

Journal télégraphique (1869-1933)
58 volumes published
Telecommunication Journal
(since 1934)
Volume 56

© ITU 1989

Published each month by the International Telecommunication Union (ITU) in separate English, French and Spanish editions

All manuscripts submitted for insertion in the "Telecommunication Journal" must reach the Editor at least 60 days before the date of publication.

CHAPTER 3

PLENIPOTENTIARY AND ADMINISTRATIVE CONFERENCES

3.1 THE LAST PLENIPOTENTIARY CONFERENCE

Nice 1989

The ITU Constitution and Convention are to be reconsidered at each Plenipotentiary Conference, the last being held at Nice in 1989.

The Final Acts of the conference include the declarations and reservations made by certain countries which signed the Final Acts. There is an optional protocol on the compulsory settlement of disputes. Also included in the Final Acts are two decisions on expenses of the ITU for the period 1990 to 1994 and the choice of members of their scale of contribution, a number of resolutions which vary in their importance and permanence, two opinions, and three recommendations.

Recent Conferences and Other Resolutions

One of the resolutions is on future conferences of the ITU, which are given below in Table 3.1 (Recent and Future Telecommunication Conferences of the International Telecommunication Union).

Some of the more important resolutions are listed in Table 3.2 (Major Resolutions of the International Telecommunication Union).

Table 3.1. Recent and Future Telecommunication Conferences of the International Telecommunication Union

- World Administrative Radio Conference for Mobile Services (Geneva, 14 September to 17 October 1987).
- Second Session of the World Administrative Radio Conference on the Use of the Geostationary-Satellite Orbit and on the Planning of Space Services Utilizing It (Geneva, 29 August to 6 October 1988).
- Second Session of the Regional Administrative Planning Conference for the Broadcasting Service in the Band 1605–1705 kHz in Region 2 (Rio de Janeiro, 23 May to 29 June 1988).
- World Administrative Telegraph and Telephone Conference (Melbourne, 28 November to 9 December 1988).
- Plenipotentiary Conference (Nice, France, 23 May to 29 June 1988).
- Second Session of the Regional Administrative Conference for the Planning of VHF/UHF Television Broadcasting in the African Broadcasting Area and Neighboring Countries (Geneva, 1963), Geneva, 13 November–8 December 1989.
- Regional Administrative Conference of the Members of the Union in the African Broadcasting Area to Abrogate the Regional Agreement for the African Broadcasting Area (Geneva, 1963), Geneva, 4–5 December 1989.
- An additional Plenipotentiary Conference if so decided by the Administrative Council at its 1991 session (Geneva, two weeks).
- World Administrative Radio Conference for Dealing with Frequency Allocations in Certain Parts of the Spectrum, taking into account the Resolutions and Recommendations of WARC HFBC-87, WARC MOB-87, and WARC ORB-88 relating to frequency allocation (Spain, first quarter of 1992, four weeks and two days).
- Regional Administrative Radio Conference to Establish Criteria for the Shared Use of the VHF and UHF Bands Allocated to Mobile Broadcasting and Fixed Services; if necessary, planning for the broadcasting service in all or part of Region 3 and countries concerned in Region 1, to be determined by the Administrative Council after consultation with members concerned.
- Plenipotentiary Conference (Japan, 1994, five weeks), to be confirmed by the Administrative Council at its 1991 session.

Table 3.2. Major Resolutions of the International Telecommunication Union[*]

Resolution PLEN/5	Interim arrangements to enable the commencement of the work of the Telecommunications Development Bureau.
Resolution PLEN/6	Convening of the Plenipotentiary Conference to consider the results of a study on structural reform.
Resolution PL-A/1	Changing telecommunication environment.
Resolution PL-B/2	Future conferences of the ITU.
Resolution PL-B/3	Establishment of a voluntary group of experts to study allocation and improved use of the radio frequency spectrum and simplification of the radio regulations.
Resolution COM 6/8	ITU regional presence.
Resolution COM 7/1	Review of the structure and functioning of the International Telecommunication Union.

Telecommunications Development Bureau

The Nice Plenipotentiary Conference inserted Article 14 in the Constitution, which sets up the Telecommunications Development Bureau. The BDT is to discharge the ITU's dual responsibility as a United Nations specialized agency within its sphere of competence and executive agency for implementing projects under the United Nations development system. The BDT is to facilitate and enhance telecommunication development by offering, organizing, and coordinating technical cooperation and assistance activities.

The BDT's specific functions are to

- raise the level of awareness of decision-makers concerning the important role of telecommunication in the national socioeconomic development program;
- promote the development, expansion, and operation of telecommunication networks and services, particularly in developing countries;

---

[*] The resolution numbers are those given by the Nice Plenipotentiary Conference (1988). Definitive numbers will be assigned when the Final Acts of the conference are published.

- enhance the growth of telecommunication through cooperation with regional telecommunication organizations and global and regional development financial institutions;
- encourage participation by industry in telecommunication development in developing countries;
- advise and execute or sponsor studies as necessary, including studies of specific projects in the field of telecommunication; and
- collaborate with the CCIs and other concerned bodies in developing a general plan for international and regional telecommunication networks.

The BDT is to work through World Development Conferences, Regional Development Conferences, and an elected director.

Standards Preeminence and Standards Cooperation

A substantial resolution which will have some effect on the way in which the ITU approaches the problem of maintaining its position in world standards is the resolution on the "Changing Telecommunication Environment."

The essence of this resolution is that the ITU is to make its role in coordinating international telecommunication more effective by

- strengthening its cooperation on subjects of mutual interest concerning telecommunication with
  —other United Nations organizations,
  —other international organizations having a specific relationship with the United Nations,
  —other multilateral organizations,
  —regional and subregional telecommunication organizations,
  —United Nations regional economic commissions,
  —regional and subregional broadcasting organizations,
  —the principal nongovernmental international organizations, and
  —professional, scientific, and academic institutions concerned with telecommunication;
- continuing and expanding initiatives to have user communities participate, where appropriate, in the formulation of international telecommunication policies and regulations.

This resolution also paid significant attention to the requirements of developing countries by continuing international initiatives to close the "telecommunications gap" between developing and developed countries.

Management Review

Of the utmost importance is the need for the ITU to continue to lead in telecommunication standards setting, a position which is increasingly under challenge because of its methods of working, membership, and resources. The ITU considered the continuing growth in the volume and complexity of the task to be performed, the changing need of the environment, the need for economy and efficiency owing to economic constraints, and

- the need for the structure, management practices, and working methods of the ITU to respond to the changes resulting from the above factors and to the increase in the demands placed upon it to keep pace with the accelerating progresss in telecommunication.

The Plenipotentiary Conference therefore resolved that a high level committee be formed to execute an in-depth review of the structure and functioning of the ITU and to recommend measures to ensure greater cost effectiveness within and between all ITU organs and activities. The committee should recommend measures to improve the structure, organization, finance, staff, procedures, and coordination, in particular by identifying and analyzing:

- options for the structure of the ITU and its permanent organs; and
- the internal management of the permanent organs, focusing on the most effective organization for work, and cost-effective and harmonized working procedures.

The study should include recommendations on improved financial management, financial transparency, and accountability. The study should also look at interaction between the permanent organs including the role of the Co-ordination Committee with a view to ensuring greater harmonization of the activities of these organs.

The Administrative Council, at an extraordinary session held in November 1989 established the committee and defined precise procedures for tasks including general guidelines to the committee on its activities. The Administrative Council is to consider whether an additional Plenipotentiary Conference ought to be called as early as possible to implement all or part of the recommendations of the study. If held, the additional plenum will be a two-week conference, probably at Geneva. Its agenda would be confined to proposed amendments to the Constitution and Convention and the election of the director of the BDT. The BDT is to operate forthwith under the aegis of the Secretary-General.

Review of Radio Regulations

Another important resolution resulted from the need to review the service definitions (Radio Regulations, Article 1), to cater for converging technologies, and to develop the basis for a future review of the Table of Frequency Allocations (Radio Regulations, Article 8) including an examination of alternatives to the way in which the radio frequency spectrum is allocated. The objective of this review would be to maximize the efficient use of the frequency spectrum to cover multifunctional radio systems and to improve the Administrative Provisions to allow more service and system sharing.

The panel of experts, set up in accordance with Resolution 68 of the Plenipotentiary Conference (Nairobi, 1982), concluded in their final report that there was a need to simplify the regulatory procedures in the Radio Regulations and the related terminology, phraseology, and mechanisms. Accordingly, the Nice Plenipotentiary Conference resolved to establish a voluntary group of experts to study allocation and improved use of the radio frequency spectrum and simplification of the Radio Regulations. This group of experts would report with recommendations to the 1992 session of the Administrative Council for utilization of the radio spectrum and report to the 1993 session of the Administrative Council with recommendations on the simplification of the Radio Regulations in general. The Administrative Council will forward these reports and their conclusions to the administrations, and will consider including them on the agenda of an appropriate World Administrative Radio Conference.

Ratification, Acceptance, and Approval

Article 40 of the Constitution, on Administrative Regulations, provides that ratification, acceptance, or approval of the Constitution and Convention constitutes consent to be bound by the Administrative Regulations adopted by competent World Administrative Conferences prior to the date of signature (30 June 1989) of the Constitution and Convention.

There also is provision for a member being deemed to have consented to be bound by revisions to the Administrative Regulations if the member fails to notify that it does not consent to be so bound.

There also were formal budgetary amounts set for annual expenses of the Administrative Council and the permanent organs, including the Telecommunications Development Bureau, World Conferences, Regional Conferences, CCIR Meetings, CCITT Meetings, and seminars.

3.2 ADMINISTRATIVE CONFERENCES

Article 9 of the International Telecommunication Constitution deals with administrative conferences. Administrative conferences of the ITU are either world or regional. The procedural provisions for administrative conferences are in Article 2 of the Convention.

Administrative conferences are normally convened to consider specific telecommunication matters. Only items included in the agenda may be discussed by such conferences. Decisions of administrative conferences must conform with the provisions of the Constitution and Convention, and such conferences in their resolutions and decisions must take into account their foreseeable financial implications.

World Administrative Conferences

The agenda of a world administrative conference may include:

- the partial revision of the administrative regulations;
- the complete revision of one or more of those regulations (an extraordinary case); and
- any other question of a worldwide character within the competence of the conference.

Regional Administrative Conferences

The agenda of a regional administrative conference may provide only for specific telecommuncation questions of a regional nature, including instructions to the International Frequency Registration Board regarding its activities with respect to the region concerned, provided that such instructions do not conflict with the interests of other regions.

The decisions of a regional administrative conference must conform with the provisions of the administrative regulations.

3.3 WORLD ADMINISTRATIVE RADIO CONFERENCE FOR MOBILE SERVICES (MOB-87)

More than 800 delegates from 108 countries together with observers representing 19 major international organizations were at the Geneva International Conference Center on 14 September 1987 to attend the World Administrative Radio Conference on Mobile Services.

The conference had the task of examining and, where appropriate, revising the provisions of the Radio Regulations governing the mobile services, mobile satellite, and radionavigation satellite, and radiodetermination satellite services with a view to the year 2000, while bearing in mind the requirements and interests of all countries, both developed or developing.

The results of the conference are contained in the Final Acts, amounting to some 480 pages. They include revisions to the Radio Regulations and are binding on the members of the ITU.

Distress and Safety at Sea

A new Chapter IX dealing with distress and safety communication in the Global Maritime Distress and Safety System (GMDSS), developed by the International Maritime Organization (IMO), was established, containing all the provisions with a specific bearing on the GMDSS. These included:

- frequencies for distress and safety communication in the GMDSS;
- operational procedures for distress and safety communication in the GMDSS; and
- alert signals.

A number of resolutions were adopted, for example, on:

- the study and implementation of a global land and maritime safety system; and
- the coordination of the use of HF Maritime Mobile Frequencies for broadcast of marine safety information on the high seas.

Aeronautical Mobile Service and Aeronautical Mobile Satellite Service

The provisions governing the Aeronautical Mobile Service and the Aeronautical Mobile Satellite Service were reexamined and modified according to the requirements of changes in particular fields.

The conference adopted two resolutions for public correspondence; one relating to the development of a global system of public correspondence with aircraft, and the other concerning the use of the frequency bands allocated to this service.

Maritime Mobile and Maritime Mobile Satellite Services

The provisions governing the Maritime Mobile and Maritime Mobile Satellite Services were updated in the light of technical progress such as increased use of radiotelephony. Procedures were established, in particular, for digital selective calling.

Landsat Mobile Service

A new definition of "Land Mobile Earth Station" was adopted together with new definitions concerning land earth stations, base earth stations, and the land and aeronautical mobile services.

3.4 WARC ON THE USE OF THE GEOSTATIONARY SATELLITE ORBIT

The First Session of the World Administrative Radio Conference on the Use of the Geostationary Satellite Orbit and the Planning of Space Services Utilizing It (ORB-85), concluded on 14 September 1985 in accordance with Resolutions 1 and 8 of the Plenipotentiary Conference, at Nairobi, 1982.

The Second Session (ORB-88) of the World Administrative Radio Conference on the Use of the Geostationary Satellite Orbit and the Planning of Space Services Utilizing It, was held in Geneva from 29 August to 6 October 1988. Some 900 delegates from 120 countries and 15 international organizations attended. The conference was chaired by Yugoslavia with the four major committees being chaired by Brazil, France, The Netherlands, and Czechoslovakia.

Background

Since the 1970s, growth in the use of the geostationary satellite orbit for communication has been such that a scarcity of orbital positions in certain sectors and frequency bands became a reality.

Many countries, especially developing countries, thought that they would be excluded from using domestic satellite communication systems because the available orbit-spectrum resource would be used up by those countries which first established satellite systems. This resulted from the virtual "first come, first served" practice, which did not allow for reservations for future use of the GSO.

As a result of an initiative by India, which at that time had problems obtaining a preferred orbital position because of INTELSAT satellites, a decision was made at a major ITU conference in 1979 to hold a World Administrative Radio Conference (WARC), which would establish in practice *equitable access* by all countries to the GSO.

The simplest method of guaranteeing access was to allocate orbital positions to all countries, irrespective of the time at which they wished to launch satellites. This is known as "*a priori* planning." This approach was strongly favored by developing countries because of its simplicity, but it has been strongly opposed by major developed countries because it ties up spectrum and orbital positions which could otherwise be used as demand dictates.

At the First Session (WARC ORB-85), a number of principles for planning were drawn up and a decision was taken to plan *a priori* a *part* of the frequency bands allocated to the fixed-satellite service and the other parts by the prevailing first-come, first-served arrangement, but with "improved procedures."

The task of the Second Session (WARC ORB-88) was to produce an allotment plan and the improved procedures for the fixed-satellite service as well as to address some other satellite communication matters, including broadcasting.

A large proportion of the conference was devoted to broadcasting satellite matters, the most important of which was the development of a feeder link (uplink) plan for the 12-GHz broadcasting satellite service (BSS) downlink allotment plan drawn up in 1977.

Outcome of the Conference

Following the conference, the developing countries were satisfied that their future access to the GSO was guaranteed in practice, and the developed countries and satellite operators did not think themselves overly constrained by the measure of inflexibility that was due to the allotment plan. Most countries welcomed the introduction of a multilateral approach to satellite system coordination. There was a general consensus that the *status quo* should remain for certain frequency bands, mainly those associated with defense systems and technological applications.

A feeder link plan for the broadcasting satellite service was produced, but there was less success with allocation of spectrum on a worldwide basis for high definition television (HDTV) and very little progress on allocating a suitable band for satellite audio broadcasting.

In summary, this Second Session of the WARC produced:

- an allotment plan for the fixed-satellite service (FSS) giving all countries a position in the GSO for 800 MHz of spectrum and a single coverage of their territory;
- a set of procedures associated with the FSS allotment plan to provide for future modifications, implementation of systems in accordance with the plan, requests for additional uses, and methods for dealing with "existing systems," which are already or planned to be operating in the same frequency band;
- regulatory provisions for the establishment of informal multilateral planning meetings (MPMs);
- simplified and improved regulatory procedures for coordinating systems outside the FSS allotment plan;
- a broadcasting-satellite feeder link plan for the 1977 12-GHz BSS plan;
- provisions for continued examination of spectrum requirements for HDTV and satellite audio broadcasting; and
- several resolutions and recommendations relating to the future work of the ITU on space communication matters, including a recommendation relating to multiservice, multiband satellite systems.

The Final Acts of the WARC ORB-88, which include partial revisions of the ITU Radio Regulations that complement the ITU International Telecommunication Convention (Nairobi, 1982), entered into force on 16 March 1990.

3.5 THE 1988 WORLD ADMINISTRATIVE TELEPHONE AND TELEGRAPH CONFERENCE

The Plenipotentiary Conference of the International Telecommunication Union (Nairobi, 1982) decided that a World Administrative Telegraph and

Telephone Conference (WATTC) should be convened immediately after the CCITT Plenary Assembly in 1988 to consider proposals for a new international regulatory framework for telecommunication services to meet the emergence and introduction of new telecommunication services. To that end, the conference through Resolution 10, instructed the CCITT to prepare proposals for submission to the Administrative Council and the IXth Plenary Assembly in 1988, to prepare the agenda for the WATTC, and to make all the necessary provisions for its organization.

The VIIIth CCITT Plenary Assembly held at Málaga-Torremolinos in October 1984, agreed to set up a Preparatory Committee for the 1988 World Administrative Telegraph and Telephone Conference (PC/WATTC-88).

CCITT Study Groups I, II, and III considered specific questions relating to their respective fields of activity and connected with Telecommunication Regulations. The main purpose of revising the regulations, which were previously revised in 1973, included allowing for the settlement of international accounts by telecommunication providers in special drawing rights (SDRs) as well as gold francs, the form of settlement now in effect. There were to be a number of other items which were also to be reviewed to bring the document in line with current technology and the regulatory environments emerging throughout the world. There were only two main services in 1973, namely telephone and telegraph services. There were in reality no specific regulatory provisions covering telex, data transmission, or any of the telematic services (that is, facsimile, teletex, videotex). The diverse range of telecommunication services now available and the number of often diametrically opposed regulatory environments throughout the world presented complex problems in the development of the regulations.

There were widely differing views regarding which services should be subject to the regulations and also which service providers should be subject to them. Telecommunication administrations that wished all services and entities providing services to be subject to the regulations had, almost without exception, both a regulatory and an operating role in their home markets. Those which had a largely deregulated market suggested that the services covered by the regulations be only those "services agreed through lateral or multinational negotiations between the specific nations involved."

Underdeveloped countries were also concerned to retain sovereign rights within their territory, being aware of the power of the more developed countries to offer services in the undeveloped country without giving it any control or compensation.

The delegations attending the earlier PC/WATTC conferences also appeared to be staffed mainly by personnel from the telecommunication administrations, rather than those responsible for the development of telecommunication and a wider national industrial policy. Input from users generally may have softened the line taken by the relevant telecommunication administrations.

Following the last meeting of PC/WATTC, informal activities ran at a greater pace than the formal meetings in an endeavor by then Secretary-General of the ITU, Mr. Richard Butler (Australia), to bring about a means of consensus which would result in implementations of the regulations that came out of WATTC-88. In a helpful and brave stance, Mr. Butler, operating at the limit of his ITU mandate, was able to draw consensus from different positions and, with the aid of Dr. Wilenski, a skilled chairman from the host country, Australia, the regulations were approved by the conference.

The conference resolution required that there be " ... established to the extent necessary ... a broad international regulatory framework." Broad international regulations were required as basic guidelines for members and not a detailed "legal" code of operating practices. The regulations needed to be general enough in their wording to take account of wide differences in national telecommunication industry structures. The regulations could not exceed the scope of the ITU Convention and were to be regarded as a suitably flexible instrument for a multilateral agreement. Where the convention did not define terms, new definitions were not automatically introduced. Without such definitions, there is naturally a greatly increased flexibility.

3.6 THE INTERNATIONAL TELECOMMUNICATION REGULATIONS

Contents

The current International Telecommunication Regulations were adopted in December 1988 by a World Administrative Telegraph and Telephone Conference held in Melbourne, following the IXth CCITT Plenary Assembly.

The table of contents and Articles 2 and 4 of the Final Acts are reproduced in Appendix 3A. They consist of the regulations, a short preamble, articles, appendices, resolutions, recommendations, and an opinion. The contents of the regulations are listed in Table 3.3 below.

Purpose and Scope of the Regulations

Article 1.1(a) states that "these regulations establish *general principles* which relate to the *provision* and *operation of telecommunications services* offered to the public as well as to the *international telecommunication transport means* used to provide such services" (emphasis added). The regulations also set rules applicable to administrations or recognized private operating agencies.

Table 3.3. Contents of the International Telecommunication Regulations

Preamble	
Article 1	Purpose and Scope of the Regulations
Article 2	Definitions
Article 3	International Network
Article 4	International Telecommunication Services
Article 5	Safety of Life and Priority of Telecommunications
Article 6	Charging and Accounting
Article 7	Suspension of Services
Article 8	Dissemination of Information
Article 9	Special Arrangements
Article 10	Final Provisions
Final Formula	
Appendix 1	General Provisions Concerning Accounting
Appendix 2	Additional Provisions Relating to Maritime Telecommunications
Appendix 3	Service and Privilege Telecommunications
Final Protocol	

We can see that there is a focus on both the means of transport and the service provided on such transport. Some tension can be observed in the paragraph because general words are used (e.g., "general principles" and "relate to") alongside more particular words (e.g., "rules applicable to").

There also is immediate recognition in Article 1.1(b) of the right of members to allow "special arrangements" under Article 9.

Article 1.3 spells out the purpose of the regulations which is to "promote the harmonious development and efficient operation of technical facilities" and also the "usefulness and availability to the public" of international telecommunication services.

The expression "the public" (which is defined in Article 1.2) is used in the sense of population, including governmental and legal bodies. This would seem to exclude trading entities and corporations operating on private networks.

Within the framework of the regulations the provision and operation of international telecommunication services in each relation is pursuant to mutual agreement between administrations or recognized private operating agencies (Article 1.5). Administrations are to comply "to the greatest extent practicable, with the relevant CCITT recommendation" (Article 1.6).

A member can, subject to national law, require that administrations and recognized private operating agencies (RPOAs) which operate in its territory and provide international telecommunication service to the public be authorized by that member (Article 1.7(a)).

Definitions and Access to Networks

Article 2 contains definitions of telecommunication, international telecommunication service, government telecommunication, service telecommunication, privileged telecommunication, international route, and relation. The latter definition is as follows.

Relation: Exchange of traffic between two terminal countries, always referring to the specific service if there is between their administrations and recognized private operating agencies

- a means for the exchange of traffic in that specific service
  —over direct circuits (direct relation), or
  —via a point of transit in a third country (indirect relation), and
- normally, the settlement of accounts.

Also defined are accounting rate, collection charge, and instructions.

Article 3 deals with the international network and focuses on the need to provide "satisfactory quality of service" and states that administrations and RPOAs are to determine by mutual agreement which international routes are to be used. Subject to international law, any user, by having access to the international network established by an administration, has the right to send traffic.

International Telecommunication Services

Article 4 deals with the need to have efficient working international telecommunication services, including ensuring a "minimum quality of service" corresponding to relevant CCITT recommendations with respect to

- access to the international network by users employing terminals,
- facilities and services available for customers' dedicated use,
- access to the public, whether or not they are subscribers to a specific telecommunication service, and
- a capability for interworking between different services.

Charging, Accounting, Safety, Suspension, and Information

Article 6 deals with collection charges, accounting rates, monetary unit, establishment of accounts and settlement of balances of account and service, and privileged telecommunication, the latter two subjects being covered in Appendices 1, 2, and 3 to the Regulations.

Article 5 deals with safety of life and priority of telecommunication, and Article 7 deals with suspension of services. Article 8 is also important, dealing with dissemination of information. This latter article provides for the Secretary-General to disseminate administrative, operational, tariff, or statistical information concerning international telecommunication routes and services.

Special Arrangements and Implementation

Under Article 9 members may enter into special arrangements on telecommunication matters which do not concern other members in general, in accordance with Article 31 of the Constitution:

> Subject to national laws, members may allow administrations and recognised private operating agencies or other organizations or persons to enter into special mutual arrangements with members, administrations or other organizations or persons that are so allowed in another country for the establishment, operation, and use of special telecommunication networks, systems and services, in order to meet specialized international telecommunications needs within and/or between the territories of the members concerned ...

This is an escape provision for the major powers which was necessary to ensure that the regulations would be adopted.

If a member makes reservations with regard to the application of one or more of the provisions in the regulations, other members and their administrations are free to disregard the provision or provisions in their relations with the member that has made such reservations and its administrations (Article 10.3).

The regulations, of which Appendices 2A, 2B, and 3A of this book form integral parts, have come into force on 1 July 1990.

There were 73 declarations by signatories to the protocol, reserving their rights to take such action as those members deem necessary which in some cases could take them outside the scope of the regulations.

APPENDIX 3A

EXCERPTS FROM THE FINAL ACTS OF THE WORLD ADMINISTRATIVE TELEGRAPH AND TELEPHONE CONFERENCE, MELBOURNE, 1988

TABLE OF CONTENTS

International Telecommunication Regulations

Article 4

International Telecommunication Services

4.1   Members shall promote the implementation of international telecommunication services and shall endeavour to make such services generally available to the public in their national network(s).

4.2   Members shall ensure that administrations* cooperate within the framework of these Regulations to provide by mutual agreement, a wide range of international telecommunication services which should conform, to the greatest extent practicable, to the relevant CCITT Recommendations.

4.3   Subject to national law, Members shall endeavour to ensure that administrations* provide and maintain, to the greatest extent practicable, a minimum quality of service corresponding to the relevant CCITT Recommendations with respect to:

   a) access to the international network by users using terminals which are permitted to be connected to the network and which do not cause harm to technical facilities and personnel;

   b) international telecommunication facilities and services available to customers for their dedicated use;

   c) at least a form of telecommunication which is reasonably accessible to the public, including those who may not be subscribers to a specific telecommunication service; and

   d) a capability for interworking between different services, as appropriate, to facilitate international communications.

Article 9

Special Arrangements

9.1   a) Pursuant to Article 31 of the International Telecommunication Convention (Nairobi, 1982), special arrangements may be entered into on telecommunication matters which do not concern Members in general. Subject to national laws, Members may allow administrations* or other organizations or persons to enter into such special mutual arrangements with Members, administrations* or other organizations or persons that are so allowed in another country for the establishment, operation, and use of special telecommunication networks, systems and services, in order to meet specialized international telecommunication needs within and/or between the territories of the Members concerned, and including, as necessary, those financial, technical, or operating conditions to be observed.

      b) Any such special arrangements should avoid technical harm to the operation of the telecommunication facilities of third countries.

9.2   Members should, where appropriate, encourage the parties to any special arrangements that are made pursuant to No. 58 to take into account relevant provisions of CCITT Recommendations.

* or recognized private operating agency(ies)

CHAPTER 4

INTERNATIONAL TELEGRAPH AND TELEPHONE CONSULTATIVE COMMITTEE (CCITT)

International Telegraph and Telephone
Consultative Committee (CCITT)
Place des Nations
CH–1211 Genève 20
Switzerland

Telephone:	41 22 730 5111
Facsimile:	41 22 733 7256
Teletex:	228-468 15100 = UIT

4.1 INTRODUCTION

The IXth Plenary Assembly, held in November 1988 at Melbourne, Australia, considered in more detail the direction and purpose of the CCITT. Although the results of its work are called "recommendations," they are effectively standards, which greatly affect the operations of private sector corporations as well as PTTs. The CCITT traditionally has concerned itself with matters relating to international telegraphy and telephony, but from the 1970s it has had an equal emphasis on national telecommunication. Telecommunication networks of all kinds (i.e., telex, telephone, digital, dedicated data, and ISDN) have been built by using CCITT standards. The CCITT is now faced with the recognition that, if it is to remain a worldwide eminent standards body, it must carefully consider what it seeks to achieve and the means by which to do so.

4.2 TERMS OF REFERENCE OF THE CCITT

Creation and Purpose

Article 7 in the Constitution of the International Telecommunication Union deals with its structure and states that the International Telegraph and Telephone Consultative Committee is a permanent organ of the ITU.

Article 13 refers to the International Consultative Committees and Article 13-1(2) states that the duties of the CCITT are to

- study technical, operating, and tariff questions and to issue recommendations on them with a view to standardizing telecommunication on a worldwide basis.

Expressly excluded are technical or operating questions relating specifically to radiocommunication, which come within the purview of the CCIR.

The CCITT must pay "due attention" to the study of questions and the formulation of recommendations directly connected with the establishment, development, and improvement of telecommunication in *developing countries.*

The CCITT must conduct its work with due consideration for the work of national and regional standardization bodies, keeping in mind the need for the ITU to maintain its preeminent position in the field of worldwide standardization for telecommunications.

Membership and Structure

Under Article 13-2, the CCITT has as members

- the administrations of all members of the ITU, and
- any recognized private operating agency (RPOA) or any scientific or industrial organization (SIO) which (with the approval of the member which has recognized it) expresses a desire to participate in the work of the CCITT.

Members of the ITU are thereby entitled to membership in the CCITT. There is only one membership fee for the ITU and no separate charge for the CCITT. RPOAs and SIOs are individually charged for their membership in the CCITT.

Article 6 of the Convention of the ITU further regulates the International Consultative Committees.

The CCITT is to work through the medium of

- the Plenary Assembly, preferably meeting every four years;
- study groups, which are set up by the Plenary Assembly to deal with the questions to be examined;
- a director, elected by the Plenipotentiary Conference; and
- a specialized secretariat, which assists the director.

Questions for the CCITT

The questions studied by the CCITT on which it shall issue recommendations are those referred to it by

- the Plenipotentiary Conference,
- an Administrative Conference,
- the Administrative Council,
- the CCIR,
- the IFRB,
- the Plenary Assembly of the CCITT, or
- at least twenty members of the ITU, in the intervals between Plenary Assemblies, when requested or approved by correspondence.

At the request of the countries concerned, the CCITT may also study and offer advice concerning their national telecommunication problems.

The Workings of the CCITT

Chapters III and IV of the Convention specify the way in which the CCITT is to operate, the relevant headings being listed in Table 4.1 below.

Under Article 16 the members of the CCITT may participate in all of its activities. Any requests from recognized private operating agencies and scientific or industrial organizations to take part in the work of the CCITT must be approved by the member under whose jurisdiction the agency is recognized.

A recognized private operating agency may act on behalf of the member that recognizes it, provided that the member informs the CCITT in each particular case that the RPOA is authorized to do so.

Table 4.1. Provisions for Administration of the CCITT in the Convention of the ITU

CHAPTER III

General Provisions Regarding International Consultative Committees

ARTICLE 16	Conditions for Participation
ARTICLE 17	Duties of the Plenary Assembly
ARTICLE 18	Meetings of the Plenary Assembly
ARTICLE 19	Languages and Right to Vote in Plenary Assemblies
ARTICLE 20	Study Groups
ARTICLE 21	Conduct of Business of Study Groups
ARTICLE 22	Duties of the Director, Specialized Secretariat
ARTICLE 23	Proposals for Administrative Conferences
ARTICLE 24	Relations of Consultative Committees Between Themselves and with International Organizations

CHAPTER IV

Rules of Procedure

ARTICLE 25	Rules of Procedure of Conferences and Other Meetings

International organizations and regional telecommunication organizations that coordinate their work with the ITU and have related activities may be admitted to participate in the work of the CCITT in an advisory capacity.

4.3 DUTIES OF THE PLENARY ASSEMBLY

Article 17 of the Convention deals with the duties of the Plenary Assembly. The Plenary Assembly is to

- consider the reports of study groups and approve, modify, or reject the draft recommendations contained in the reports;
- take note of the amended or new recommendations which have been approved between Plenary Assemblies;
- consider existing questions as to whether their studies ought to be continued, and prepare a list of new questions to be studied (which should be completed in twice the interval between two Plenary Assemblies);
- approve the program of work arising from the consideration of questions, determine the order of questions to be studied according to their

importance, priority, and urgency, bearing in mind the need to keep the demands of the resources of the ITU to a minimum;

- decide, in light of the approved program of work, whether existing study groups should be maintained or dissolved, and whether new study groups should be set up;
- allocate to study groups the questions to be studied;
- consider and approve a report of the director on the activities of the CCITT since the last meeting of the Plenary Assembly;
- approve, if appropriate, for submission to the Administrative Council, the estimate of the financial needs of the CCITT up to the next Plenary Assembly; and
- consider the reports of the World Plan Committee and any other matters deemed necessary.

When adopting resolutions and decisions, the Plenary Assembly should take into account the foreseeable financial implications and should try to avoid adopting resolutions and decisions which might give rise to expenditure in excess of the upper limits on credits laid down by the Plenipotentiary Conference.

Article 18 deals with meetings of the Plenary Assembly and Article 19 deals with the right to vote in Plenary Assemblies. Article 20 covers study groups and Article 21 concerns the conduct of the business of study groups.

During the Plenary Assembly, the work associated with the organization, working methods, and work program of the study groups, finance, and technical assistance is generally organized into five committees. The committees' recommendations are then presented to a plenary meeting for approval. Plenary meetings are also scheduled to consider the reports of the study groups, special autonomous groups (GAS), and plan committees. At the beginning of a Plenary Assembly, a heads of delegation meeting decides the program of work.

The five committees are

Committee A	Organization and Working Methods of the CCITT
Committee B	CCITT Work Program
Committee C	Budget Control
Committee D	CCITT Technical Assistance
Editorial Committee	

4.4 STUDY GROUPS ESTABLISHED AT THE PLENARY ASSEMBLY

Questions and Participants

The Plenary Assembly is a nontechnical meeting of CCITT member organizations, which sets up and maintains, as necessary, study groups to deal with the questions to be studied. The administrations, RPOAs, SIOs, international organizations, and regional telecommunication organizations that desire to take part in the work of the study group, give their names either at the meeting of the Plenary Assembly or at a later date to the Director of the CCITT.

The Plenary Assembly normally appoints a chairman and one vice-chairman of each study group. If the workload requires, however, additional vice-chairmen may be appointed. In appointing chairmen and vice-chairmen, particular consideration is given to the requirements of competence, equitable geographical distribution, and the need to promote more efficient participation by the developing countries.

Conduct of Business of Study Groups

Study groups conduct their work as far as possible by correspondence. The Plenary Assembly, however, may give directives concerning the convening of any meetings of the study groups that appear necessary to deal with large groups of questions. As a general rule, study groups hold no more than two meetings between sessions of the Plenary Assembly, including the final meetings held before that assembly.

Where necessary, the Plenary Assembly or the CCITT may set up joint working parties for the study of questions requiring the participation of experts from several study groups.

The director sends the final reports of the study groups to participating administrations, recognized private operating agencies, and, as occasion may demand, to international organizations and regional telecommunication organizations that have participated. The reports must be received at least one month before the date of the next meeting of the Plenary Assembly. If there is no report on the question, it does not appear on the agenda for the meeting of the Plenary Assembly.

4.5 STUDY GROUPS FOR THE 1988–1992 STUDY PERIOD

Given below in Table 4.2 are the study groups for the 1988 to 1992 study period, with their "mission statements."

Table 4.2. CCITT Study Groups and Committees for 1988–1992

STUDY GROUPS

Study Group I — Services

Responsible for questions relating to service definitions, service operation, principles of service interworking, and subscriber quality of service. Work embraces consideration of proposals from other Study Groups on both the definition of bearer services and the technical aspects of services development.

Study Group II — Network operation

Responsible for questions relating to ISDN and telephone network operation, routing, numbering, network management, and service quality of networks (traffic engineering, operational performance, and service measurements).

Study Group III — Tariff and accounting principles

Responsible for questions relating to tariff and accounting principles for services studied by the CCITT.

Study Group IV — Maintenance

Responsible for questions relating to maintenance of services and networks (including their constituent parts such as circuits, signalling systems, *et cetera*) as well as the use and application of specific maintenance mechanisms provided by other Study Groups. This includes the maintenance of digital networks including ISDN.

Study Group V — Protection against electromagnetic effects

Responsible for questions relating to the protection of telecommunication plant and equipment from dangers and disturbance of electromagnetic origin.

Study Group VI — Outside plant

Responsible for questions relating to outside plant including the construction, installation, jointing, terminating, protection from corrosion and other forms of damage, and associated structures for all types of cable for public telecommunication.

Study Group VII — Data communications networks

Responsible for questions relating to dedicated data networks, message handling systems, directory systems, and the overall responsibility for the reference model of Open Systems Interconnection for CCITT applications.

Study Group VIII — Terminals for telematics services

Responsible for questions relating to terminals for telematics services such as facsimile, teletex, videography, and telewriting including the higher level protocols relating to terminals for telematics services and document architecture.

Study Group IX — Telegraph networks and telegraph terminal equipment

Responsible for questions relating to telegraph transmission and related terminal equipment including telegraph, telex, and gentex networks.

Table 4.2. CCITT Study Groups and Committees for 1988–1992 (Continued.)

Study Group X — Languages for telecommunication applications
Responsible for questions relating to technical languages for telecommunication applications.

Study Group XI — Switching and signalling
Responsible for questions relating to ISDN and telephone network switching and signalling.

Study Group XII — Transmission performance of telephone networks and terminals
Responsible for questions concerning the end-to-end transmission performance and related transmission planning implications as applied to telephone services on the Public Switched Telephone Network (PSTN) and to other services utilizing voice band transmission connections or channels. This work includes the transmission aspects of all signals as routinely carried on the PSTN, e.g., speech, in-band signalling, and voice band data. This work also includes speech quality aspects of ISDN.

Study Group XV — Transmission systems and equipment
Responsible for questions concerning transmission systems and equipment including speech coding.

Study Group XVII — Data transmission over the telephone network
Responsible for questions relating to data transmission over circuits and networks that are accessed via an analogue interface. Additionally responsible for subject matter relating to the application of modems and terminal adaptors on ISDN, interworking between data terminals using modems on the PSTN and data terminals on an ISDN.

Study Group XVIII — ISDN
Responsible for questions concerning ISDN and related network aspects of services as well as general network aspects. Has overall responsibility for the continuing studies of ISDN taking into account the functional responsibilities of other study groups.

Terminology

REGIONAL TARIFF GROUPS OF STUDY GROUP III

GR TAF	Tariffs (Africa)
GR TAL	Tariffs (Latin America)
GR TAS	Tariffs (Asia and Oceania)
GR TEUREM	Tariffs (Europe and the Mediterranean Basin)

JOINT STUDY GROUPS

CMTT	CCIR-CCITT Joint Study Group	Television and Sound Transmission

Table 4.2. CCITT Study Groups and Committees for 1988–1992 (Continued.)

PLAN COMMITTEES

	Designation	*Title*
PLAN MONDIAL	World Plan Committee	General Plan for the Development of the International Telecommunication Network
PLAN AF	Plan Committee for Africa	General Plan for the Development of the Regional Telecommunication Network in Africa
PLAN AL	Plan Committee for Latin America	General Plan for the Development of the Regional Telecommunication Network in Latin America
PLAN AS	Plan Committee for Asia and Oceania	General Plan for the Development of the Regional Telecommunication Network in Asia and Oceania
PLAN EU	Plan Committee for Europe and the Mediterranean Basin	General Plan for the Development of the Regional Telecommunication Network in Europe and the Mediterranean Basin

SPECIAL AUTONOMOUS GROUPS

GAS 7	Rural Telecommunication	
GAS 9	Economic and Technical Aspects of Transition from an Analogue to a Digital Telecommunication Network	(Case Study of a Global Network)
GAS 12	Strategy for the Introduction of New Non-voice Telecommunication Services in Developing Countries	

4.6 RULES OF PROCEDURE OF THE CCITT

Rules Adopted by the Plenary Assembly

In addition to the rules of procedure of the CCITT appearing in the International Telecommunication Convention, the CCITT through its Resolution No. 1 (reviewed and revised, as necessary, every four years) has its own rules of procedure for organizational arrangements of meetings, documentation, logistics, *et cetera*.

These additional rules of procedure were updated in 1988 at Melbourne and are published in Volume I of the *Blue Book* and in outline are:

I   Plenary Assembly

II   The Director

III Study Groups and other groups

    1 Classification of Study Groups
    2 Classification of Other Groups
    3 Meetings outside Geneva
    4 Participation in Meetings
    5 Frequency of Meetings
    6 Preparation of Studies and Meetings
    7 Conduct of Meetings
    8 Use of Special Rapporteurs
    9 Preparation of Reports, Recommendations, and New Questions
  10 Final Meetings of Study Groups

IV Submission and Processing of Contributions

    1 Submission of Contributions
    2 Processing of Contributions.

Further details on the rules of procedure for the study groups can be found in Appendix 4A (CCITT Rules of Procedure).

4.7 CCITT RECOMMENDATIONS IN THE RED BOOK AND BLUE BOOK

The recommendations made by the study groups must be approved either at the Plenary Assembly of the CCITT or between Plenary Assemblies by agreed procedures before they formally become "recommendations." These recommendations are then published in ten volumes which, for the 1984 to 1988 period, are known as the *Red Book* and are published with bright red covers. The recommendations were published as the *Yellow Book* for the VIIth Plenary Assembly and would be published as the *Blue Book* following the IXth Plenary Assembly for the 1988 to 1992 period. The CCIR has always had its recommendations published in similar volumes, called the *Green Book,* as they are light green in color.

4.7 ISSUES AT THE 1988 PLENARY

Special Group S was created by the VIIIth Plenary Assembly of the CCITT to study the future evolution of the CCITT study group structure. Special Group S, among other things, proposed

- moving to a functional organization based on subject matter, and
- centralizing and reinforcing work on services concentrating on the task of service definition in Study Group 1.

With the globalization of telecommunication, manufacturers and network operators need agreed international standards. Special Group S proposals were really "too little, too late" and much greater change is required for the CCITT to remain a force in telecommunication standards setting.

The two working methods that have evolved are meetings of working groups of a study group at frequent intervals and less frequent meetings of full working groups, with some formation of smaller rapporteur groups that do the technical work, meeting at convenient locations.

The director of the CCITT proposed "complementary proposals," namely,

- a minimum period of eight months between formal meetings of a study group or its working parties,
- formal agreement to "interregnum" meetings between final study group meetings and the beginning of a new study period,
- much greater use of special rapporteurs, including appointment of a rapporteur per question, establishing rapporteur groups, and encouraging these groups to hold separate meetings,
- new document-handling rules, and

- other changes reflecting working methods currently being adopted, including encouragement of the greater use of accelerated approval of Recommendations.

The director's proposals should improve the productivity of the CCITT and also serve the working convenience of the CCITT Secretariat and ITU common services. If most of the meetings are held outside Geneva, the rapporteurs rather than the CCITT provide facilities, and there is an extra week to copy submitted documents, there will be much smaller use of CCITT resources and the major players (who will provide the rapporteurs) will provide more resources. This will result in an increase in power of the major players who can attend all meetings and thus may not be attractive to the majority of members of the ITU. As decisions taken by the rapporteur groups will still need complete endorsement by full meetings, some of the advantages may not be realized.

CCITT faces differing requirements of developed countries, on the one hand, whose telecommunication administrations have an increasingly commercial outlook and developing countries, on the other, who have relied to an extent on outside assistance to develop basic network services.

CCITT and similar organizations exist primarily to establish and update international standards, and thus increasingly to promote the development of international markets for telecommunication equipment and services. While development assistance will not be ignored in this scenario, it will have an increasingly commercial element; that is, aid will be directed to projects that also provide commercial gain to the donor.

The best assistance that CCITT can give to developing countries is the timely production of international standards, as is required to allow their networks to evolve in an ordered manner with standardized and compatible interfaces and equipment. The CCITT will need to guard against too much use of its resources to serve the specific interests of developing countries. Also necessary to consider is the CCITT's role in technical assistance when juxtaposed with the Center for Telecommunications Development (CTD) and the recent Telecommunications Development Bureau (BDT).

APPENDIX 4A

CCITT RULES OF PROCEDURE

Rules Adopted by the Plenary Assembly

In addition to the rules of procedure of the CCITT appearing in the International Telecommunication Constitution and Convention, the CCITT by its first resolution at the VIIIth Plenary Assembly at Málaga-Torremolinos in October 1984, resolved to amplify the general regulations of the Nairobi Convention 1982. These were amended at the IXth Plenary Assembly in November 1988 at Melbourne, Australia.

The current text of Resolution No. 1 is published in Volume I of the Blue Book for the last Plenary Assembly in 1988 at Melbourne and in outline contains

I   Plenary Assembly (PA)

II   The Director

III Study Groups and Other Groups

   1 Classification of study groups

   2 Classification of other groups

   3 Meetings outside Geneva

   4 Participation in meetings

   5 Frequency of meetings

   6 Preparation of studies and meetings

   7 Conduct of meetings

   8 Use of Special Rapporteurs

   9 Preparation of Reports, Recommendations and new Questions

   10 Final meetings of study groups

IV Submission and Processing of Contributions

   1 Submission of Contributions

   2 Processing of Contributions.

Preparation for the Plenary Assembly by Committees

The Director of the CCITT sends an invitation to participate in the Plenary Assembly (PA) to all members of the Union and to recognized private operating agencies (RPOAs) which are members of the CCITT.

The Director must be advised of the names of delegates of Administrations and representatives of RPOAs who will attend meetings of the PA. Those persons who are invited to attend the PA in advisory capacity must notify the names of their observers.

Prior to the official opening of the PA the heads of delegation shall meet to prepare the program of work of the PA for submission to the PA at its first meeting, to designate persons who will be proposed as vice-chairmen and chairman of the PA and to determine the committees to be proposed to the PA for establishment. The committees which are normally proposed are

- Committee on the Organization and Working Methods of the CCITT (Committee A),
- Committee on the CCITT Work Program (Committee B),
- Budget Control Committee (Committee C),
- Technical Assistance Committee (Committee D),
- the Editorial Committee.

Committee B considers questions for study or further study, checks for duplication and assembles related questions and submits a report to the PA on

- the wording of the questions proposed for their allocation of questions to study groups, and
- the degree of priority on each question.

The Technical Assistance Committee defines the policy to be adopted by the PA for dealing with subjects concerning the planning, establishment and management of National Telecommunication Networks which are of general interest to all country members of the ITU. This committee makes the preliminary examination of the report of the Special Autonomous Study Groups (GAS).

Any other PA committee may be set up if the Heads of Delegation deem it advisable.

Proceedings of the Plenary Assembly and Recommendations

The PA sets up all necessary committees, the study groups and other groups as appropriate, and their respective chairman and vice-chairman.

The PA examines the final reports of the study groups, and the report of the Director, approves recommendations and notes the list of questions appearing in these reports.

The Committee on the Organisational and Working Methods of the CCITT meets to prepare proposals on the organization of the work of the CCITT and the Budget Control Committee approves the budget of the current PA and prepares a report on an estimate of the financial needs of the CCITT until the next PA.

The PA appoints the chairmen and vice-chairmen of study groups, plans committees and any other groups for a new study period and the Committee on the CCITT Work Program prepares a draft program of work.

On the proposal of the Committee on the CCITT Work Program the PA shall

- prepare the list and text of questions set for study or further study until the next PA,
- allocate these questions to study groups and other groups,
- decide when a question concerns several study groups, whether to set up a joint working party comprising members of the study group concerned, or to entrust the study to a single study group, the necessary coordination being effected within the framework of national organizations, and
- where appropriate set up coordinating groups for each family of study groups comprising the chairman and vice-chairman of the study groups in the family concerned.

Recommendations prepared by study groups and put to the vote during a PA are approved if they obtain a majority of votes. The minutes of the PA give the result of the vote without listing the delegations which voted for or against unless the delegation expressly asks for its vote to be mentioned. The result of any vote when it is not unanimous is recorded by indicating the majority and minority, both with abstentions.

Study Groups and Other Groups

A clear description of the work area of each study group should be approved by the Plenary Assembly.

To facilitate their work, study groups may set up working parties for the preparatory study of some of the questions assigned to them by the PA. A joint working party set up with the approval of the PA submits draft recommendations to the controlling study group designated by the PA which established it.

Study groups may meet outside Geneva if invited to do so by Administrations or RPOAs of countries that are members of the Union and if the holding of a meeting outside Geneva is desirable. Invitations to hold meetings away from Geneva may be accepted by the Director of the CCITT if the chairmen concerned agree and if the budgetary resources of the CCITT permit (see CCITT Opinion 1).

Administrations, RPOAs and scientific or industrial organizations shall be represented in the study groups and other groups in whose work they wish to take part, by participants appointed by name and chosen by them as experts qualified to investigate technically and economically satisfactory solutions to the questions under study.

The final meetings of study groups at the end in each study period shall be not less than four months before the beginning of the PA, to ensure the preparation of full and homogenous final reports and to give delegates to the PA an opportunity to study them thoroughly. Study groups normally meet once in the period between the end of the PA and the final meeting before the following PA, however additional meetings may be held with the approval of the Director to discuss questions which cannot be resolved by correspondence. In particular, such extra meetings could facilitate the introduction of new recommendations using the accelerated procedure for approval.

A Study Group may request meetings to be conducted on ongoing activities under the existing organization of the Study Group in the time period between the final meeting and its first meeting in the new study period in consultation with the Director of the CCITT. Such activities are listed in the report of its final meeting.

Meetings of study groups should, if possible, be arranged so as to

- enable participating bodies to send one delegate or representative to cover several meetings,

- enable the exchange of information which is required without delay,
- enable specialists in the same or related subjects to have direct contact with each other for the benefit of their Administrations, and
- enable the specialists concerned to avoid leaving their home countries too often.

The work program and the time table of meetings is prepared and communicated to participating bodies one year in advance of the meetings to enable these bodies time to study problems and submit contributions within the prescribed time limits and to enable the CCITT Secretariat to distribute the contributions. If study group chairmen and delegates have the opportunity to consider the contributions in advance, the meetings should be more efficient and shorter.

All Administrations, RPOAs, scientific, industrial and international organizations are advised of the dates of the final study group meetings at least three months in advance so that they may submit draft questions for study in the next study period.

Preparation of Studies and Meetings

At the beginning of each study period an organization proposal and an action plan for the study period should be prepared by each chairman with the help of the CCITT Secretariat. The action plan is implemented according to contributions received from the members of the CCITT and the view expressed by participants in the meetings.

Two months before each meeting, a collective letter with a draft work plan and a list of the questions to be examined is prepared by the CCITT Secretariat with the help of the chairman.

A meeting is cancelled if an insufficient number of normal contributions have been submitted. If it appears from the contributions received there is agreement among Administrations on the reply to a question (or part of a question), the chairman with the assistance of the CCITT Secretariat may submit a draft reply, possibly accompanied by a draft recommendation.

If the contributions reveal divergent proposals or points of view, the Secretariat may prepare a summary of the different positions on the question and the stage reached in the study.

Conduct of Meetings

The chairman may decide that there should be no discussion on questions on which an insufficient number of contributions have been received. Questions which have not illicted any contributions should not be placed on the agenda of the meeting and at the end of the study period should be deleted from the list of questions to be studied.

Study groups and working parties may set up small working teams during their meetings, to study allocated questions.

Working Parties

A study group may entrust questions to an *ad hoc* working party for preliminary study. It may, by agreement with another study group or other study groups, entrust an *ad hoc* joint working party with the study of questions of common interest. In such cases the controlling study group has final responsibility for the study. The contributions used as a basis for discussion in the *ad hoc* joint working party shall be sent exclusively to the members registered in the *ad hoc* joint working party. Only the final reports of *ad hoc* joint working parties shall be sent to all members of the study groups concerned.

Working parties or sub-working parties should be set up by study groups only after thorough consideration of the questions. Proliferation of working parties and sub-groups should be avoided as far as possible.

The meetings of regional tariff groups is in principle limited to delegates and representatives of Administrations and RPOAs. However, each regional tariff group may invite other participants to attend all or part of a meeting.

The format for liaison statements between study group and working parties includes

- the origin of the liaison document
- the nature of information and the goal (e.g. action, comment, information, etc.)

Use of Special Rapporteurs and Liaison Representatives

The study of questions by correspondence is encouraged as far as practicable, at least during the early stages of study. For this purpose, a study group or working party may instruct a Special Rapporteur appointed by the study group or working party and working alone or in collaboration with others, to make a preliminary analysis of the more complex questions or to prepare a draft reply to a question for submission to the study group or working party.

The Special Rapporteur may, when working with a group of collaborators, choose whatever method of work he considers appropriate, e.g., correspondence or meetings of the group.

The following are examples of ways in which a Special Rapporteur may work

- analyse the contents of contributions distributed,
- study a single question requiring detailed consideration before decisions can be taken, and
- call an informal meeting of his collaborators when the study cannot progress further by correspondence alone.

He submits a report on the results of his work to the next meeting of the study group or working party.

The Special Rapporteur, after consulting with his collaborators on the necessity for a meeting and the availability of contributions, may call a meeting of his collaborators subject to the agreement of the chairman of the study group or working party chairman and the Director of the CCITT. Where close coordination is required between Special Rapporteurs and several study groups who are dealing with closely related questions, a meeting of the Special Rapporteurs may be called subject to agreement of the chairman of the study groups concerned and the Director of the CCITT. The Special Rapporteur is responsible for all necessary arrangements for meetings, ensuring adequate notice is given to all collaborators who are encouraged to participate.

When an area of study requires coordination between several study groups, Liaison Representatives are nominated. The Special Rapporteurs themselves may be nominated as the Liaison Representatives to one or more study groups, or several Liaison Representatives for a question within a study group may be appointed depending on the number of other study groups with which liaison is required. It is the responsibility of the Liaison Representatives together with the CCITT Secretariat, to ensure effective liaison with the involved study groups based on specific terms of reference and working methods determined by the study group.

Progress reports with proposed draft recommendations to be translated should be received by the Director of the CCITT two months before the next meeting of the parent study group of the working party. Additional reports and liaison documents to be made available for a meeting of a study group or a working party should be received at the CCITT Secretariat at least seven working days before the meeting.

Preparation of Reports, Recommendations and New Questions

A report on the work done during a meeting is prepared by the Secretariat of the CCITT. This report summarises the results of the meeting, the agreements reached and identifies the issues for further study. The number of annexes to the report is kept to a strict minimum.

The report consists of two parts, namely

PART I

Organization of work, main results, directive for future work, planned meetings of working parties, sub-working parties and Rapporteur Groups, condensed liaison statements endorsed at the study group level, references to contributions or documents, issued during a meeting.

PART II

Draft Recommendations or modified Recommendations accepted by the meeting as mature enough to be considered by the Members.

The study group or working party may arrange for delegates to draft some parts of the report. The report should be submitted for approval before the end of the meeting, or later to the chairman for his approval.

Any meeting of a working party is the subject of a report. In the absence of a report prepared by the CCITT Secretariat, the chairman of the meeting is responsible for preparing the report, a copy of which is sent to the CCITT Secretariat, for circulation as a white document (normal contribution) to all members of the study group in question.

Draft recommendations and resolutions are prepared by the chairman with the assistance of the CCITT Secretariat or drafting groups and must be approved in final form before the end of the meeting.

The accelerated procedure for the approval of recommendations should be used when it appears to a study group that a draft has reached an adequate state of completion and agreement. The time remaining before the next Plenary Assembly should be taken into account.

Draft new recommendations should not be submitted to the PA for approval while the texts are of a provisional nature.

The preliminary examination of draft new questions is supervised by the chairman who applies stringently the criteria of CCITT Opinion No. 3. When questions arise during the interval between PAs, they may be studied when requested or approved by correspondence by at least 20 members of the Union.

Final Meetings of Study Groups

All study groups meet sufficiently in advance of the PA for the final report of each study group to reach Administrations at least one month before the PA. Study groups may appoint one or two members to prepare a text summing up the final results of the work.

The final report of each study group which is forwarded to the PA, is the responsibility of the study group chairman and shall include

- a short, comprehensive summary of the results achieved in the study period, in particular, which new and amended draft recommendations have been prepared, identifying the scope, application and importance of these draft recommendations,
- the final text of new draft recommendations and of draft amendments to existing recommendations,
- the text of provisional recommendations to be adopted by the PA, approved by correspondence under Resolution No.2,
- the list of questions proposed to the PA for the following study period.

Submission and Processing of Contributions

Contributions to current studies are sent, by official letter to the Director of the CCITT. Contributions received at least two months before a meeting are published in the normal way.

Contributions received by the Director less than two months but not less than seven working days before the date set for the opening of a meeting shall be published as "Delayed Contributions" in the form in which they are received and shall be distributed at the beginning of the meeting to only the participants present. Administrations, RPOAs, scientific and industrial organizations should advise the CCITT Secretariat about any forthcoming "Delayed Contributions" and their contents at least two months before the meeting.

The CCITT Secretariat does not reissue delayed contributions as normal contributions unless otherwise decided by the study group or working party, and delayed contributions are not included in reports as annexes.

Contributions are submitted to a single study group. However, if an administration submits a contribution which it believes is of interest to several study groups it identifies the study group primarily concerned and the other study groups involved. The contribution is issued in full to the study group primarily concerned and a single sheet giving the title of the contribution, its source and a summary of its contents is issued to the other study groups. The single sheet is numbered in the series of contributions of each study group to which it is issued.

Extracts from reports of other study group meetings or from reports of chairmen, special rapporteurs or drafting groups received less than two months before the meeting are published as temporary documents.

CHAPTER 5

INTERNATIONAL
RADIO CONSULTATIVE COMMITTEE

International Radio Consultative Committee (CCIR)
Place des Nations
CH–1211 Genève 20
Switzerland

Telephone:	41 22 730 51 11
Facsimile:	41 22 733 72 56
Teletex:	228-468 15100 = UIT

5.1 TERMS OF REFERENCE OF THE CCIR

Article 5 in the Constitution of the International Telecommunication Union deals with its structure, and states that the International Radio Consultative Committee is a permanent organ of the ITU.

Article 11 covers International Consultative Committees and states that the CCIR's duties under the Convention are to

- study technical and operating questions relating specifically to radiocommunication without a limit of frequency range, and to issue recommendations on them, with a view to standardizing telecommunications on a worldwide basis; these studies will not generally address economic questions, except where technical options are compared.

The CCIR is to pay "due attention" to the study of questions and the formulation of recommendations directly connected with the establishment, development, and improvement of telecommunication in developing countries, both regionally and internationally.

The objectives of the CCIR as outlined by its XIVth Plenary Assembly are to

- provide the technical bases for use by administrative radio conferences and radiocommunication services for efficient utilization of the radio-frequency spectrum and the geostationary satellite orbit, bearing in mind the need of the various radio services,
- recommend performance standards for radio systems and technical arrangements which ensure their effective and compatible interworking in international telecommunication, and
- collect, exchange, analyze, and disseminate technical information resulting from studies by the CCIR and other information available for the development, planning, and operation of radio systems, including any special measures required to facilitate the use of such information in developing countries.

5.2 PARTICIPATION IN THE CCIR

The provisions of the Constitution of the International Telecommunication Union govern participation in the work of the CCIR.

Categories of Participants

The CCIR has as members

- administrations of members of the ITU, and
- any recognized private operating agencies and scientific or industrial organisation, which, with the approval of the member concerned, expresses a desire to participate in the work of the CCIR.

Members of the ITU are automatically members of CCIR, however recognized private operating agencies and scientific or industrial organizations need to request membership subject to approval by the administration concerned.

Also involved in CCIR work in an advisory capacity are

- international and regional telecommunication organizations that coordinate their work with the ITU, subject to approval by a majority of its Members.

All organizations participating in the work of the CCIR are listed in Volume XIV-1 of the Recommendations and Reports of the CCIR (1986).

Financial Conditions for Participation

Administrations, by virtue of their membership in the ITU, contribute to its budget, including the CCIR. All other participants make financial contributions according to the contributory class which they have chosen. The scale ranges from 40 units down to 1/16 of a unit. The contributory unit is fixed by the Administrative Council of the ITU and at the 37th session of the council (1982) was 34,000 Swiss francs per annum, subject to review at subsequent sessions.

International organizations may be exempted by the Administrative Council on condition of reciprocity.

Procedure to Become a Participant

There are no formal requirements for participation by administrations. Recognized private operating agencies and scientific or industrial organizations (with the approval of the ITU member that recognizes the agency or organization) expressing a desire to participate may also take part in the CCIR's work. The Secretary-General informs all members and the director of the CCIR of such new participants.

A request by an international or regional organization to participate in the CCIR is received by the Secretary-General, who then informs all members by telegram and invites them to respond. The request is granted on his receipt within one month of a majority of favorable replies.

All participants, except those admitted on the basis of reciprocity, indicate in their application the contributory class in which they wish to be placed and the specific study groups in which they wish to participate, giving details of address and mailing particulars for study group documents.

5.3 STRUCTURE OF THE CCIR

The work of the CCIR is done at a number of levels. The highest forum through which the work of the CCIR passes is the Plenary Assembly of the CCIR. This assembly meets every four years to consider, amend, and approve the recommendations made by the study groups (at present there are thirteen) that have been set up by it to conduct the practical work of the CCIR. Following approval by the Plenary Assembly, the work is published.

The study groups, one of which is a joint study group with the CCITT, are set out in Table 5.1 below.

The research and work conducted by each study group during successive four-year study periods is initiated by a series of questions to be answered, relating to the particular subject matter allocated to the group. The study groups of the 1986–1990 study period meet on two occasions over the four years between Plenary Assemblies to discuss and settle draft recommendations and reports that constitute the answers to their allocated questions.

Table 5.1. CCIR Study Groups for the 1986–1990 Study Period

1	Spectrum Utilization and Monitoring
2	Space Research and Radioastronomy Services
3	Fixed Service at Frequencies below about 30 MHz
4	Fixed Satellite Service
5	Propagation in Nonionized Media
6	Propagation in Ionized Media
7	Standard Frequencies and Time Signals
8	Mobile, Radiodetermination, and Amateur Services
9	Fixed Service Using Radio-Relay Systems
10	Broadcasting Service (Sound)
11	Broadcasting Service (Television)
12	CMTT-CCIR-CCITT Joint Study Group for Television and Sound Transmission
13	CMV (until 1988, a Joint Study Group with CCITT): Vocabulary

The questions and study programs are usually decided by the study groups themselves and then adopted by the Plenary Assembly at its next meeting. Questions may also be referred to a study group by the ITU Plenipotentiary Conference, an Administrative Conference, the Administrative Council, the CCITT, the IFRB, and the agreement of at least twenty members of the ITU. Study programs are established to deal with specifics under each question.

The basic working units of the study groups are the Interim Working Parties (IWPs), the formation of which must be approved by the Plenary Assembly. The working parties are allocated specific questions to research. The work is done by technical experts in the relevant field meeting or corresponding regularly. These specialists exchange ideas, viewpoints, and the results of their research on the particular aspect of radiocommunication on which they are employed and report their findings to the study group.

The reports and recommendations of the CCIR generally have no legal force and are voluntary in application. The reports and recommendations are of importance in regulating radiocommunication matters, however, because they are vital to the operation of radiocommunication systems and equipment throughout the world. The reports and recommendations are frequently followed by the world's radiocommunication operators and equipment manufacturers. The reports and recommendations are also used by Administrative Radio Conferences as the basis of Radio Regulations, which have the force of treaties among nations attending the conferences.

5.4 WORK OF THE CCIR

Recent expansion of its jurisdiction into satellite radiocommunication and the higher frequencies of the radio spectrum is reflected in its recent and continuing work, which includes

- effective and efficient use of the geostationary orbit,
- antenna patterns for earth stations and satellites,
- dissemination of standard frequencies and time signals by satellites,
- use of data relay satellites,
- design and manufacture of earth stations to access information from satellites,
- modulation and coding techniques,
- interference criteria,
- introduction of digital transmission for use of frequencies above 10 GHz,
- development of standards for digital coding and enhanced quality high definition television systems,
- the effect of rain and other atmospheric conditions on radiocommunication, and
- radio-frequency sharing and efficient use of spectrum.

Some countries form National Study Groups (NSGs) which correspond with the CCIR study groups. National Preparatory Groups (NPGs) are sometimes formed for the specific purpose of preparing a brief for the national delegation to the Administrative Radio Conferences. Unlike the NSGs, which remain as continuing establishments, an NPG finishes its work following completion of the delegation report.

5.5 ADMINISTRATIVE RADIO CONFERENCES

The CCIR prepares detailed technical reports for specialized Administrative Radio Conferences. Study groups are asked to prepare these conference reports using their research and recommendations to the Plenary Assembly as the basis for the reports.

Administrative Radio Conferences may occur at the rate of up to two per year. They are formal meetings attended by most members of the ITU. Considerable discussion occurs among 150 representatives to reach agreement on the many minute details involved in radiocommunication. Reports prepared by representatives are considered, as are CCIR reports.

Agreement results in the adoption by participants of treaty instruments known as *Final Acts* that make up part of the Radio Regulations. These regulations carry legal force among participant nations. The ever increasing number of regulations means an ever increasing number of details with which to comply. An obvious disadvantage of the many regulations promulgated by the conference is a lack of flexibility when the existing technology becomes dated. However, recent conferences have attempted to provide a suitable method of modification to accommodate fast changing technology.

Suggestions have been made that Administrative Conferences and the resultant regulations should be almost wholly abandoned, and all material that would otherwise be considered by conferences transferred to study groups of the CCIR and treated as recommendations to be approved by the Plenary Assembly.

5.6 RECENT AND CURRENT ACTIVITIES

The Second Session of the WARC for the Planning of the High Frequency (HF) Bands Allocated to the Broadcasting Service (WARC HFBC-87) was held in Geneva from 2 February to 8 March 1987. The CCIR's report to this conference provided relevant information such as refinement of field strength calculation and the allowance for interference between double-sideband and single-sideband emissions.

The Second Session of the WARC for Mobile Services (WARC MOB-87) was held in Geneva from 14 September to 16 October 1987. One of the main items for MOB-87 was the introduction of a Global Maritime Distress and Safety System (GMDSS). The GMDSS will represent a significant step forward for maritime services. MOB-87 also made provision for domestic mobile communication via satellite. The CCIR's report to this conference provided relevant technical information as well.

Study Group 11 (Television Broadcasting) will give further consideration to a proposed new world standard for high definition television, having approximately twice the resolution of conventional television, a wide screen 16:9 aspect ratio, and intended for studio production and international exchange of program material.

In September 1988, the Second Session of the WARC on the Use of the Geostationary Orbit and the Planning of Space Services Utilizing It (WARC ORB-88) was held. The ORB-88 conference has had a major effect on the regulatory provisions for planning and implementation of future satellite services. The CCIR's report covered such matters as allotment planning, improved regulatory procedures, satellite audio broadcasting, and satellite HDTV transmission.

5.7 THE CCIR VOLUMES

The Green Book comprises the recommendations approved by the Plenary Assembly for each study group, details of the proceedings of the Plenary Assembly including resolutions, reports, and the new study groups, their questions, and study programs for the ensuing four years.

Volume XIV-1 contains a useful summary of the Plenary Assembly, and the Table of Contents from Volume XIV-1, and the plan of Volumes I to XIV of the Green Book are given in Appendix 5A.

APPENDIX 5A

CONTENTS OF CCITT GREEEN BOOK VOLUME XIV-1 STRUCTURE OF THE CCIR FROM RECOMMENDATIONS AND REPORTS OF THE CCIR AND PLAN OF VOLUMES I TO XIV, 1986

TABLE OF CONTENTS OF VOLUME XIV-1

PLAN OF VOLUMES I TO XIV
XVITH PLENARY ASSEMBLY OF THE CCIR

(Dubrovnik, 1986)

VOLUME I	Spectrum utilization and monitoring.
VOLUME II	Space research and radioastronomy.
VOLUME III	Fixed service at frequencies below about 30 MHz.
VOLUME IV-1	Fixed-satellite service.
VOLUMES IV/IX-2	Frequency sharing and coordination between systems in the fixed-satellite service and radio-relay systems.
VOLUME V	Propagation in non-ionized media.
VOLUME VI	Propagation in ionized media.
VOLUME VII	Standard frequencies and time signals.
VOLUME VIII-1	Land mobile service. Amateur service. Amateur-satellite service.
VOLUME VIII-2	Maritime mobile service.
VOLUME VIII-3	Mobile satellite services (aeronautical, land, maritime, mobile and radiodetermination). Aeronautical mobile service.
VOLUME IX-1	Fixed service using radio-relay systems.
VOLUME X-1	Broadcasting service (sound).
VOLUMES X/XI-2	Broadcasting-satellite service (sound and television).
VOLUMES X/XI-3	Sound and television recording.
VOLUME XI-1	Broadcasting service (television).
VOLUME XII	Transmission of sound broadcasting and television signals over long distances (CMTT).
VOLUME XIII	Vocabulary (CMV).
VOLUME XIV-1	Information concerning the XVIth Plenary Assembly: Minutes of the Plenary Sessions. Administrative texts. Structure of the CCIR. Lists of CCIR texts.
VOLUME XIV-2	Alphabetical index of technical terms appearing in Volumes I to XIII.

CHAPTER 6

INTERNATIONAL ORGANIZATION FOR STANDARDIZATION

International Organization for Standardization (ISO)
1, rue de Varembé
CH–1211 Genève 20
Switzerland

Telephone:	41 22 734 12 40
Facsimile:	41 22 733 34 30

6.1 INTRODUCTION

The International Organization for Standardization (ISO) was created in February 1947 as a result of a conference attended by representatives of twenty-five national standards bodies and held in London in October 1946.

The first constitution and rules of procedure of the ISO were drafted at the London conference and formally adopted with ratification being received from fifteen of the national standards bodies present at the conference. The constitution was last revised in 1985.

6.2 ISO TODAY

The ISO comprises the national standards bodies of 88 countries. There are 73 member bodies and 15 correspondent members. ISO's technical work is carried out by some 2,500 technical bodies (167 technical committees, 651 subcommittees and 1,659 working groups and 26 *ad hoc* study groups). The secretariats of the technical committees and subcommittees are distributed among 32 member bodies.

The work of ISO has resulted, to date (*circa* 1989), in the publication of 7,225 ISO international standards, and 2,480 draft proposals for international standards are registered at the ISO Central Secretariat. A further 2,033

---

new work items appear on the programs of work of the technical committees, but have not yet reached the draft proposal stage.

Some 430 international organizations have been granted liaison status with ISO technical committees and subcommittees. The services provided to the technical secretariats by the member bodies equal a full-time staff of about 400 persons. The Central Secretariat has a staff of 145 persons drawn from 24 countries.

In 1988, ISO published 527 international standards, representing a total number of 6,394 pages, 625 draft international standards were submitted to the ISO member bodies for voting, and a total of 862 technical meetings were held in 25 countries.

6.3 THE OBJECTS OF THE ISO

The object of the ISO is to promote the development of standardization and related activities in the world with a view to facilitating international exchange of goods and services, and developing cooperation in intellectual, scientific, technological, and economic activities.

To achieve this the ISO may

- take action to facilitate worldwide harmonization of standards and related activities,
- develop and issue international standards and take action for their worldwide implementation,
- arrange for exchange of information regarding work of its member bodies and technical committees, and
- cooperate with other international organizations interested in related matters, particularly by undertaking at their request studies pertaining to standardization projects.

6.4 MEMBERSHIP OF THE ISO

The member bodies are those national standards bodies that are most representative of standardization in their respective countries, which have been admitted to the ISO in accordance with its rules of procedure. Only one body in each country can be admitted to membership.

National bodies interested in standardization in countries without a member body may be registered as correspondent members without voting rights, in accordance with the procedure defined by the ISO's council.

6.5 ORGANIZATION OF THE ISO

Voting and Decisions

A decision by vote of the member bodies, either in the General Assembly or by mail ballot, constitutes a *decision* of the ISO. Decisions in the ISO on technical matters are only recommendations to the membership, each member remaining free either to follow or not to follow them. In principle, international standards are

- industrywide,
- reached by consensus, and
- voluntary.

The Organs of the ISO

The organs of the organization are the General Assembly, the council, the Executive Board, the Technical Board, technical committees, and the Central Secretariat.

General Assembly

The General Assembly is constituted by a meeting of delegates nominated by the member bodies. Each member body may nominate not more than three official delegates, although the latter may be accompanied by observers. The General Assembly, which usually meets once every three years, met most recently at Tokyo in 1985 and at Prague in 1988.

The Council

The operations of the ISO are administered in accordance with the policy established by the ISO, that is, the General Assembly in session. Between the General Assembly meetings, the council, consisting of the president, vice-president, treasurer, and representatives of eighteen member bodies elected for a three-year term, administers the operations of the ISO. The council meets at least once a year.

The council reports on the activities of the ISO to the member bodies each year and to the General Assembly at each of its sessions. To function effectively, the council creates committees entrusted with studies in a particular field. Such committees report to the council each year on the matters referred to them.

The ISO may cooperate with other international organizations partially or wholly interested in standardization or related activities. The conditions of cooperation have been established by the council. The funds of the ISO are derived from the dues and contributions of the members and from the sale of publications. The acceptance of funds from other sources is at the discretion of the council.

The council may undertake work within the ISO's objectives at the request of, or on behalf of, other international organizations and may accept reimbursement for expenses incurred. The council submits to the member bodies a budget for the ensuing year.

ISO Council Resolutions

Council resolutions adopted between its first meeting in 1947 and the 41st meeting in 1987 are published and comprise all decisions that deal with matters of principle, or may be considered of current interest. Exceptions are council resolutions on ISO membership and on the creation and scope of ISO technical committees, the latter information being given in the ISO *Memento*.

The Executive Board

The Executive Board consists of the vice-president and nine other persons appointed by the council. The council determines the terms of reference of the Executive Board and may delegate to the board such functions and duties as are necessary. The Executive Board reports to the council.

Technical Board

The Technical Board consists of a Chairman and nine other persons appointed by the council. The council determines the terms of reference of the Technical Board and may delegate to the board such functions and duties as are necessary. The Technical Board reports to the council.

The key tasks delgated to the Technical Board are

- appointment of technical committee chairs;
- allocation or reallocation of technical committee secretariats;
- approval of titles, scope, and programs of technical committee work;
- assignment of priorities, if necessary, to particular items of technical work;
- coordination of technical work;

- reviewing the need for, and planning of, work in new technical fields;
- maintenance of the joint IEC-ISO directives and other rules for technical work; and
- consideration of matters of principle raised by national bodies and appeals of decisions on new work, proposals, committee drafts, and draft international standards.

Technical Advisory Groups

Technical Advisory Groups (TAGs) are established by the Technical Board to advise it on matters of basic, sectoral and cross-sectoral coordination, coherent planning, and the need for new work. These Technical Advisory Groups are

TAG 1	Chemical and Physico-chemical Test Methods and Methods of Analysis
TAG 2	Metals
TAG 4	Metrology
TAG 5	Fire Tests
TAG 6	Medical Equipment (Joint TAG with IEC)
TAG 7	Information Technology Applications
TAG 8	Building
TAG 9	Distribution of Goods
TAG 10	Image Technology
TAG 11	Safety

Technical Committees

The technical committees (TCs) are established by the council and work under its authority. Each member body interested in a subject for which a technical committee has been authorized has the right to be represented on that committee. The scope and program of work for each technical committee is approved by the Technical Board on behalf of the council.

6.6 THE WORKING OF TECHNICAL COMMITTEES

Technical committees are listed in full in the ISO *Memento*. The basic information is provided for each ISO technical committee, their subcommittees (SCs), and working groups (WGs). SCs and WGs may be established by technical committees to cover different aspects of their work. ISO technical committees are assigned numbers in chronological order, beginning with TC 1 established in 1947, and continuing to TC 196 established in 1989. The year of establishment appears in italics under each TC number published in the *Memento*.

When a technical committee is dissolved, its number is not reallocated. Twenty-eight technical committees have been dissolved. Each technical committee and subcommittee has a secretariat assigned to an ISO member body. The abbreviations of all member bodies are listed in the *Memento*. When an individual has been designated as secretary of a technical committee, his or her name appears in brackets under the secretariat acronym.

Conveners of Working Groups

A working group does not have a secretariat, but rather has an individual convener appointed by the parent committee to which he or she will report. The *Memento* lists the member body through which the convener can be reached.

The chair of a technical committee or subcommittee need not be a national of the country holding the secretariat. Chairs are appointed by the Technical Board on behalf of ISO council and subcommittee chairs by the parent technical committee. The *Memento* shows the year at the end of which the term of office expires. Appointments made up to 1989 were typically for a three-year term. From 1990, appointments will normally be for six years.

Central Secretariat

The Central Secretariat comprises the Secretary-General and such staff as the ISO requires, and is located in the same place as the ISO, Geneva, Switzerland.

Each member of the ISO respects the international character of the responsibilities of the Central Secretariat staff and cannot seek to influence them in the discharge of their responsibilities.

Publications and Documents of the Organization

The ISO may issue

- international standards;
- international standardized profiles;
- technical reports, guides, and documents for study purposes;
- reports on progress of work within the technical committees;

- minutes of meetings of the General Assembly, the council, and the technical committees; and
- various publications and documents related to the work of the ISO and its members.

The major documents published by the ISO in addition to international standards follow.

ISO Memento

This publication is issued annually and contains information in English and French (there is a separate Russian version) on the scope of responsibility, organizational structure, and secretariats for each ISO technical committee. The *Memento* also contains general information on the organization and administration of the work of ISO and a list of members.

ISO Catalogue

The annual catalogue lists all published ISO standards and is updated quarterly by cumulative supplements.

KWIC Index

The fourth edition of the ISO *KWIC* index of international standards has been issued in 1989. Prefaced by the Director-General of the General Agreement on Tariffs and Trade (GATT), the index covers nearly 12,000 international standards produced by ISO, IEC, and 27 other international organizations.

ISO Standards Handbooks

These handbooks are an easy means for referring to international standards, as each of them reproduces the full text of ISO standards in a given field.

Selective Lists

Selective lists of international standards on a given subject are selected from the ISO database and arranged according to combinations of fields and groups in the ISO Catalogue. The lists are produced by the ISO Central Secretariat upon request and are fully up-to-date at the time of printing.

ISO Technical Program

The ISO Technical Program is published twice yearly and includes all proposed standards being considered or being amended. A description of the stages in the standards-making process is detailed in section 60.

ISO Bulletin

The monthly *Bulletin* provides standardization news, a calendar of ISO meetings, a list of new draft ISO standards, and newly published standards.

Other Publications

An ISO booklet, "Access to Standards Information — How to Inquire or Be Informed About Standards and Technical Information Available Worldwide," addressed more particularly to outside users of information centers, was issued in 1986.

6.7 RULES OF PROCEDURE AND MEMBERSHIP

The rules of procedure deal with membership, the General Assembly, the council, consultation of member bodies, technical committees, the appointment of the president, vice-president, treasurer, and secretary-general, finances, rights of reproduction, voting by mail ballots, and elections.

The tenure of office of the member bodies elected to the council commences on 1 January and expires on 31 December three years following the commencement of the term. Each member body retiring from the council is eligible for reelection at the end of the term. One-third of the member bodies to be represented on the council are elected each year.

The election of the member bodies to be represented on the council occurs in the General Assembly or, in the years in which no General Assembly is held, by mail ballot. The council determines the dates and places of its meetings.

Notice of a meeting, its agenda, and the necessary relevant documents are circulated by the secretary-general to members of the council at least four months prior to the meeting. When council meets, each member body elected to the council may be represented by not more than three persons. In meetings of council, decisions are taken by a majority vote of the member bodies elected to the council and represented at the meeting.

The Formation of Technical Committees

A request for undertaking the study of a technical subject may be initiated by one or more member bodies, a technical committee, a council committee, the secretary-general, or an organization outside the ISO.

Such a request should be accompanied by a statement setting forth the scope envisaged and the reasons that, in the opinion of the originator, warrant undertaking the project.

The secretary-general after having further clarified the request, if necessary, enquires of the member bodies whether they are in favor of the formation of a technical committee to study the subject proposed and, if the reply is in the affirmative, whether they elect to participate actively in the work of the technical committee established or merely to follow its activity.

If the majority of the member bodies voting are in favor of a technical committee being authorized and at least five members are willing to take an active part in the work, the creation of such a committee may be authorized provided that the council is convinced of the international importance of the work. No technical committee may exceed the scope of the work outlined for it unless such extension is approved by the council.

A member body is appointed by the Technical Board on behalf of the council to act as secretariat of a technical committee. If working as secretariat of a technical committee, the member body must maintain strict neutrality and distinguish sharply between proposals that it makes as a member and those in its capacity as secretariat. The secretariat of a technical committee is responsible to the council for its work and correspondence.

The Creation of Standards

The ISO and IEC jointly issue directives for the work of the technical committees. Unless otherwise provided by the council, all matters are decided by a majority vote of the member bodies participating in the work of the technical committee, either at meetings or by correspondence. The step toward an international standard is the committee draft (CD), formerly draft proposal (DP), a document circulated for study within the technical committee. The stages of standards making and the stage numbers are set out in Table 6.1.

When agreement is reached with the committee, the CD is sent to the Central Secretariat for registration as a draft international standard (DIS); the DIS is then circulated to all member bodies for voting. If it is approved by at least two-thirds of members participating in the committee and 75% of all members, the DIS is accepted as an international standard. Normally, the fundamental technical issues are resolved at technical committee level.

The ISO has established the general rule that all ISO standards are to be reviewed at not more than five-year intervals. On occasion, a standard may need to be revised earlier.

Table 6.1. The Stages of Standards Making and the Stage Numbers

STAGE \ SUB-STAGE	Action by office of CEO **0**	Approval procedure initiated **1**	Comments/ voting summary circulated **2**	Action on voting results **3**	Receipt for further action **4**	Approval procedure initiated **5**	Comments/ voting summary circulated **6**	Action on voting results **7**	Approved for higher stage **8**	Referral/ revision/ cancellation **9**
Preliminary stage **0** *(See note 2 opposite)*					0.4 Proposal for new project received	0.5 Proposal for new project under review	0.6 Review summary circulated		0.8 Approval to ballot proposal for new project	
Proposal stage **1**	1.0 Proposal for new project registered	1.1 New project ballot initiated	1.2 Voting summary circulated						1.8 New project approved	1.9 Proposal returned to submitter for further definition *(See note 2 opposite)*
Preparatory stage **2**	2.0 Project registered in TC or SC work programme	2.1 Working draft (WD) study initiated	2.2 Comments summary circulated						2.8 WD approved for registration as CD	
Committee stage **3**	3.0 Committee draft (CD) registered	3.1 CD study/ballot initiated	3.2 Comments/ voting summary circulated						3.8 CD approved for registration as DIS	
Approval stage **4**	4.0 DIS registered	4.1 DIS ballot initiated	4.2 Voting summary dispatched [Next step: 4.3, 4.8 or 4.9] ‒ ‒ ‒ ‒ 6 months ballot procedure	4.3 Full report circulated in case of decision for 2 months ballot	4.4 Amended draft received at office of CEO	4.5 DIS ballot initiated	4.6 Voting summary dispatched [Next step: 4.8 or 4.9] ‒ ‒ ‒ ‒ 2 months ballot procedure		4.8 Full report circulated and DIS approved for publication as International Standard	4.9 DIS referred back to TC or SC
Publication stage **5**	5.0 International Standard under publication		5.2 Proof sent to TC or SC secretariat	5.3 International Standard published		5.5 International Standard under periodical review	5.6 Review summary dispatched	5.7 International Standard confirmed		5.9 International Standard under revision
[Reserved for future use] **6**										
[Reserved for future use] **7**										
[Reserved for future use] **8**										
Withdrawal **9**	9.0 Withdrawal of International Standard proposed by TC or SC	9.1 Withdrawal ballot initiated	9.2 Voting summary dispatched	9.3 New project rejected or project deleted				9.7 International Standard replaced by new edition	9.8 International Standard replaced by another International Standard	9.9 International Standard withdrawn without being replaced

6.8 COLLABORATION WITH IEC AND OTHER INTERNATIONAL ORGANIZATIONS

International Electrotechnical Commission

For many years, the ISO has had a close working relationship with the International Electrotechnical Commission (IEC), a body which is responsible for standards in electrical and electronic engineering. Recent years have witnessed a strengthening of ISO-IEC collaboration through the continuing activity of joint groups, such as the ISO-IEC Joint Technical Programming Committee (JTPC) and the ISO-IEC Advisory Board on Technological Trends (ABTT), and the establishment of the first ISO-IEC Joint Technical Committee (JTC 1: "Information Technology") by the ISO and IEC councils on the JTPC's recommendation. The ISO-IEC Joint Technical Programming Committee was set up for technical coordination in all cases where one of the two parent organizations feels a need for joint planning to avoid or eliminate overlapping in technical work.

Day-to-day collaboration on specific technical matters occurs primarily through formal liaisons established between the technical bodies of ISO and IEC, and involves exchanges of information and requests for comments on drafts.

Invitations to participate in the meetings of each other's technical bodies are issued by ISO and IEC in all cases where there is common interest.

Relations with Other International Organizations

ISO work is of interest to many other international organizations. Some of these make a direct technical contribution to the preparation of ISO standards. Others, particularly intergovernmental organizations, contribute to the implementation of ISO standards, for example, by utilizing them in the framework of intergovernmental agreements.

International organizations may be granted "liaison status" with ISO technical committees and subcommittees. Liaison status comprises two categories: A (effective contribution to the work) and B (to be kept informed only). Liaison A gives the right to submit papers, attend meetings, and participate in discussions. At present, some 440 international organizations are in liaison with ISO technical committees and subcommittees. The publication, *ISO Liaisons,* which is available from the Central Secretariat or ISO member bodies, gives full information on the technical nature of these liaison arrangements.

ISO has consultative status (Category 1) with the UN Economic and Social Council (ECOSOC), and equivalent status with nearly all other bodies and specialized agencies of the UN.

6.9 COUNCIL COMMITTEES OF THE ISO

There are a number of so-called "council committees," as listed in Table 6.2, reporting annually to the council on policy matters within their relevant fields.

Table 6.2. Council Committees of the ISO

- Committee on Conformity Assessment (CASCO)
- Committee on Consumer Policy (COPOLCO)
- Development Committee (DEVCO)
- Committee on Information (INFCO)
- Committee on Reference Materials (REMCO)
- Committee on Standardization Principles (STACO)

The council committees which are most relevant to telecommunication and information processing are CASCO and INFCO.

6.10 CASCO

Introduction

CASCO is the ISO Council Committee on Conformity Assessment. Its precedessor was the ISO Council Committee on Certification (CERTICO), which was set up by the ISO council in 1970 and whose name and terms of reference were changed in 1985 to reflect more explicitly the field of application of the committee's work.

Participation is open to all ISO member bodies and to the IEC. ISO corresponding members may become observers if they so wish. The secretariat duties of CASCO and its working groups are entrusted to the ISO Central Secretariat. The CASCO chairman is appointed by ISO council for a three-year term of office, renewable only once.

The IEC participates in CASCO in accordance with the ISO-IEC Agreement of 1976. This cooperation takes the form of representation at meetings of CASCO and its working groups, parallel circulation of basic draft documents of interest to both organizations, and consultations at national level between member bodies of the two organizations.

Objectives

The objectives of CASCO are to

- study the means of assessing the conformity of products, processes, services, and quality systems to appropriate standards or other technical specifications;
- prepare international guides relating to the testing, inspection, and certification of products, processes, and services, and to the assessment of quality systems, testing laboratories, inspection bodies, certification bodies, and their operation and acceptance; and
- promote mutual recognition and acceptance of national and regional conformity assessment systems, and the appropriate use of international standards for testing, inspection, certification, assessment, and related purposes.

Work Program

The current work program, as approved by ISO council, includes the review of existing ISO-IEC guides on the competence of calibration and testing laboratories, the revision of the booklet on certification, the development of guidelines on the presentation of inspection results, ongoing advice to the technical committees, and preparation of ISO-IEC guides in response to requests arising from the International Laboratory Accreditation Conference (ILAC), as well as a study on the way to promote recognition and acceptance of certification systems established on the basis of ISO-IEC guidelines.

The current Working Groups are

- CASCO WG4—Presentation of Inspection Results;
- CASCO WG5—Definitions; and
- CASCO WG6—Revision of ISO-IEC Guides 25 and 38 on the competence of laboratories.

6.11 INFCO-ISONET

The ISO Information Network (ISONET) coordinates and systematizes the exchange of information on standards and standard-type documents both internationally and nationally, by linking the information centers of the ISONET members and that of the Central Secretariat.

INFCO acts as the general assembly of ISONET and in this capacity promotes the objectives stated in the ISONET Constitution. In addition, as an ISO council committee, INFCO advises the council on policy matters related to the compilation, storage, retrieval, application, dissemination, and promotion of scientific and technical information on standards, technical regulations, and related matters. The ISONET management board is the executive organ of ISONET. The chairman of the board is appointed by INFCO, subject to confirmation by council.

The basic documentation for ISONET comprises

- *ISONET Constitution,* third edition, 1985. Trilingual: English, French, Russian. This publication also includes the conditions for participation in ISONET as a national member, an associate member, or an international affiliate.
- *ISONET Manual,* first edition, 1985. Separate English and French editions. The *Manual* defines elements of information which should be considered for inclusion in records of standards or technical regulations in the information systems of standards bodies, and gives guidance on how to represent the various elements in a record.
- *ISONET Directory,* second edition, 1987. Bilingual. The *Directory* contains the list of addresses and services offered by ISONET members, associate members, and international affiliates.
- *ISONET Guide,* second edition, 1988. Separate English and French editions. The *Guide* gives guidelines for the operation of an ISONET information center and recommended procedures for information interchange.

In addition, the *ISONET Communiqué* is issued two or three times a year. It contains news from ISONET members and the secretariat. The Communiqué is intended for circulation to all staff working in ISONET information centers.

6.12 STACO

The Committee on Standardization Principles (STACO) was set up to

- provide an international forum for the exchange of views and sharing of experience relating to fundamental aspects of standardization, and

- conduct those studies that are entrusted to it by the ISO council on subjects related to the basic principles of standardization, including methodology and terminology.

This committee has 38 participating countries and 24 observer countries.

6.13 ISO PLANNING

The ISO, like the ITU, has been compelled to consider revamping its organizational structure and standards-making procedure to adapt to the demand for standards, particularly in the telecommunication and information technology areas.

Accordingly, the ISO has been planning

- methods by which international standards may be produced and revised far more quickly than has previously been the case (consolidated in harmonized ISO-IEC directives for technical work);
- the setting of priorities by technical committees formulating their own strategic plans;
- more emphasis on the ISO technical boards establishment of an advisory board on technological trends, and a long-range planning survey (Delphi Survey);
- greater cooperation with other standards-making bodies and user bodies; and
- restructuring the ISO to optimize its organizational structure, particularly with emphasis on quality control.

CHAPTER 7

THE INTERNATIONAL ELECTROTECHNICAL COMMISSION (IEC)

International Electrotechnical Commission (IEC)
3, rue de Varembé
CH-1211 Genève 20
Switzerland

Telephone: 41 22 34 01 50
Facsimile: 41 22 33 38 43

7.1 INTRODUCTION

The International Electrotechnical Commission, also known as Commission Electrotechnique Internationale, was set up to facilitate the coordination and unification of national electrotechnical standards. The IEC was formed in Saint Louis, Missouri, in 1904 and had its 1989 general meeting in Brighton, UK.

The IEC publishes international standards, which, although they are not binding on member organizations, are expected to be followed by members when drawing up their national standards so as to unify all national standards.

Activities

The IEC's work covers almost all spheres of electrotechnology, including power, electronics, telecommunication, and nuclear energy. This work can be divided into two categories:

- improving understanding between electrical engineers of all countries by drafting common means of expression, unification of nomenclature, agreements on quantities and units, their symbols, and abbreviations, and graphical symbols for diagrams; and

- standardization of electrical equipment proper, involving the study of problems of electrical properties and materials used in electrical equipment, standardization of guarantees to be given for certain equipment as to the characteristics, methods of test, quality, safety, and dimensions controlling interchangeability of machines and electrical equipment.

7.2 MEMBERSHIP AND NATIONAL COMMITTEES

Any country desiring to participate in the work of the IEC forms a national committee for its own country, but any country may use any other appropriate designation. Members of the IEC are those national committees that have agreed to abide by its statutes and rules of procedure.

There is only one national committee for each country. When this committee has been accepted as a member, it is known as "*the* national committee" of the country concerned. National committees have one vote in the IEC Council or by correspondence to participate in a decision of the IEC.

The national committees are required to be as representative as possible of all electrical interests in the country concerned: manufacturers, users, governmental authorities, and teaching and professional bodies. The 40 countries which are at present members include Australia, Brazil, Canada, People's Republic of China, Denmark, Finland, France, Federal Republic of Germany, India, Israel, Italy, Japan, Republic of Korea, Democratic Peoples' Republic of Korea, The Netherlands, Norway, Sweden, Switzerland, Union of Soviet Socialist Republics, United Kingdom, and United States.

7.3 ORGANIZATION OF THE IEC AND THE COUNCIL

The IEC has a Council, a Committee of Action, technical committees, a Central Office, and three advisory committees. The IEC is administered by a Council on which all the national committees are represented. The organization of the IEC is shown in Table 7.1 below.

The operations of the IEC are administered by the council, which consists of the president of the IEC, presidents of the national committees, past presidents and vice presidents of the IEC (without vote), treasurer, and general secretary (both *ex officio* without vote). The minutes of meetings of the council are sent to all national committees. The council meets at least once a year.

The IEC derives its income principally from annual dues paid by its national committees to the central office and from the sale of publications.

Table 7.1. Organization of IEC

COUNCIL

COUNCIL
COMMITTEE OF ACTION
TECHNICAL COMMITTEES   ADVISORY COMMITTEES
                                                            (ACET, ACES, and ACEC)
SUBCOMMITTEES   CENTRAL OFFICE
WORKING GROUPS

ACET   Advisory Committee on Electronics and Telecommunications
ACES   Advisory Committee on Safety
ACEC   Advisory Committee on Electromagnetic Compatibility

The IEC publishes draft standards proposed by the appropriate technical committee after they are approved by a vote under the *six months rule.* Amendments are approved under the *two months procedure.*

7.4 THE COMMITTEE OF ACTION

The Committee of Action consists of the president of the IEC and twelve members nominated by national committees and elected for a six-year period by the council. National committees that have made proposals to the Committee of Action are invited by the president to send a representative to participate in discussion of their proposals.

The Committee of Action deals with the tasks delegated to it by the council and takes any action considered necessary to ensure the satisfactory operation of the technical work of the IEC. The Committe of Action reports all its decisions to the council and meets at least once a year; more frequent meetings, however, may be called.

7.5 TECHNICAL COMMITTEES

Establishment of Technical Committees

The technical work of the IEC is done by technical committees, each dealing with a particular subject. The scope of the technical committee is clearly defined at the time of its formation and is subject to the approval of the council. The Committee of Action and at least one-third of the national committees must also approve any proposal for setting up a technical committee.

Each national committee interested in the subject for which a technical committee has been established has the right to participate in all the activities of that committee. Each technical committee has a chairman and a secretariat appointed by the council. The IEC may publish technical reports, approved under the same rules, pending the issue of a "standard" on the subject of the report.

Rules of Procedure

The proposed standard to be put to a vote is circulated by the central office to all the national committees. The national committees vote on the proposal within six months of the date of dispatch of the communication by the central office. The proposal is adopted unless one-fifth or more of the members have cast a negative vote.

The two months procedure is applied only to ensure the approval of amendments to a text adopted under the six months rule if the chairman of the technical committee considers that the use of such a procedure is likely to increase the number of national committees that will ratify the draft as amended.

Reports are issued as approved by the national committees and these reports contain the results of the technical work by the IEC. *Standards* are intended to serve as the basis of national standards for the different member countries. A report may be published as an intermediate step.

Relevant Technical Committees

There are 81 technical committees. Major technical committees and subcommittees from a telecommunication point of view are shown in Table 7.2.

7.6 RECENT EVENTS

In 1986 the relationship between the IEC and the ISO was further developed, with the councils of both organizations agreeing to circulate proposals for adopting harmonized working procedures and common IEC-ISO drafting rules. In 1987 the first joint technical committee with ISO was formed, namely, JTC 1: Information Technology, comprising IEC TC 83: Information Technology Equipment, IEC SC 47B: Microprocessor Systems, and ISO TC 97: Information Processing Systems. This committee met in November 1987 at Tokyo to discuss the organization and working methods that would meet the needs and challenges of international information technology (IT) standardization. JTC-1 has since been very active.

Table 7.2. IEC Technical Committees Relevant to Telecommunication

TC	Subject	SC	Subject
1	Terminology		
3	Documentation and Graphical Symbols		
		3A	Graphical Symbols for Diagrams
		3B	Documentation
		3C	Graphical Symbols for Use on Equipment
12	Radiocommunications	12A	Receiving Equipment
		12B	Safety
		12C	Transmitting Equipment
		12D	Aerials (Antennas)
		12E	Radio Relay and Fixed Satellite Communications Systems
		12F	Equipment Used in the Mobile Services
		12G	Cabled Distributions Systems
25	Quantities and Units, and Their Letter Symbols		
46	Cables, Wires, and Waveguides for Telecommunication Equipment		
		46A	RF Cables
		46B	Waveguides and Their Accessories
		46C	LF Cables and Wires
		46D	Connectors for RF Cables
47	Semiconductor Devices	47A	Integrated Circuits
48	Electromechanical Components for Electronic Equipment	48B	Connectors
		48C	Switches
		48D	Mechanical Structures for Electronic Equipment
52	Printed Circuits		
56	Reliability and Maintainability		
60	Recording	60A	Sound Recording
		60B	Video Recording
74	Safety of Information Technology Equipment Including Electrical Business and Telecommunication Equipment		
76	Laser Equipment		

Table 7.2. IEC Technical Committees Relevant to Telecommunication (Continued.)

TC	Subject	SC	Subject
84	Equipment and Systems in the Field of Audio, Video, and Audiovisual Engineering		
86	Fiber Optics	86A	Fibers and Cables
		86B	Fiber Optic Interconnecting Devices and Passive Components
CISPR	International Special Committee on Radio Interference		X,B,C,D,E,F, and G
JTC 1	ISO-IEC Joint Technical Committee for Information Technologies		
		47B	Microprocessor Systems
		83	Information Technology Equipment

The IEC took a significant step forward in its efforts to accelerate the standards-making process for electrotechnology products by adopting a new method called the *technical trend document,* which allows the development of a "prestandard" with a limited life in cases where the preparation of a complete IEC standard and its final approval entails the risk of being out of date before publication.

In 1988 the IEC published 245 international standards comprising 4,670 pages of engineering specifications. Some 367 draft standards were circulated to national committees for approval and more than 45,000 pages of working documents, the basis of future IEC standards, were circulated within the organization. In addition, more than 2,000 subjects were under consideration as future IEC standards.

Both IEC and ISO are represented jointly in the GATT group on multilateral trade negotiation (MTN) agreements and arrangements in the ongoing Uruguay Round. IEC also cooperated with GATT by updating the list of relevant standards for GATT's "notified products list." Both the IEC and GATT promote international trade in electrotechnical equipment.

7.7 TESTING AND QUALITY

The IECEE

The IEC System for Conformity Testing to Standards for Safety of Electrical Equipment (IECEE) brought to a total of 7,000 the number of certificates issued to date and increased its membership to 30 countries, 20 of which issue and recognize the IECEE's test certificates. The IECEE agrees to prepare guidelines for inspection of product conformity to type-tested specimens. To improve access to the services of the system, the manufacturer is allowed to have its product certified by any IECEE national certification board that adheres to the relevant standards.

The IECQ

The IEC Quality Assessment System for Electronic Components (IECQ), approves new manufacturers, listing more than 100 approved manufacturers and 40 approved distributors. Thermistors and television receiver picture tubes have been added to the range of component types approved under IECQ.

In 1986 the IEC published four new standards covering the fiber optics sector, including generic specifications for fiber optic switches, branching devices, and connectors for optical fibers and cables, as well as a standard for transmission star couplers of fiber optic branching devices.

7.8 RECENT CHANGES

India became a certifying member of IECQ in 1988, bringing the total number of certifying members to 14. Electronics manufacturers began for the first time to report sales of IECQ-approved components, and the number of manufacturers approved under the system increased to 122.

The IECQ also signed an agreement with CODUS Ltd. in the United Kingdom to set up a database giving worldwide access to on-line information about the IECQ qualified products list.

Some nineteen new or revised standards were published in the field of audio-visual and broadcasting transmission, including the parameters that affect interchangeability of specific prerecorded optical reflective video disks, specifications for minimum performance requirements for headphones and earphones, a standard covering the mechanical characteristics of nonperforated magnetic tape, a revised version of its basic standard for audio cassettes, an updated version of its world standard for cabled distribution systems, a revision of the time and control code for video tape recordings, and a standard for the minimum performance requirements of high-fidelity combination equipment, loudspeakers, and systems, as well as FM radio tuners.

For broadcast transmitters, the IEC published a new standard covering methods of measurement for television transmitters and transponders employing insertion test signals, another dealing with interface standards for transmitter systems using dedicated interconnections, and two new standards specifying methods of measurement for equipment used in terrestrial radio relay systems.

CHAPTER 8

JOINT TECHNICAL COMMITTEE—INFORMATION TECHNOLOGY (JTC 1)

Secretariat ISO/IEC JTC 1
American National Standards Institute
1430 Broadway
New York, New York 10018

Telephone:	212 641 4934
Facsimile:	212 302 1286

8.1 INTRODUCTION

The ISO and the IEC have collaborated for many years and therefore have formed the ISO-IEC Joint Technical Programming Committee (JTPC). This committee has the task of avoiding or eliminating possible or existing overlap in technical work relating to the activities of new and established technical committees.

As both the ISO and the IEC had a significant volume of work in information technology, the JTPC requested the advice of a group of experts in information technology drawn from ISO and IEC, called the Information Technology Management Group (ITMG). The ITMG made recommendations to JTPC at a joint JTPC-ITMG meeting in January 1987, on seeking the respective councils' endorsement of the establishment of a Joint Technical Committee, titled "Information Technology" (JTC 1), integrating the existing ISO TC 97 and its subcommittees and IEC TC 83 and subcommittee 47B, each of which was dealing with generic information technology work. The JTPC made this proposal, which was then approved by the ISO and IEC councils. ISO and IEC technical committees working in information technology applications (e.g., banking, industrial automation, electronic data interchange) as well as the CCITT are associated with the new JTC 1 through normal liaison arrangements.

8.2 JTC 1 SCOPE AND SUBCOMMITTEES

The scope of JTC 1 is standardization in the field of information technology. JTC 1 has the task of managing the development of generic information technology standards, coordinating with ISO and IEC technical committees dealing with other aspects of information technology, maintaining appropriate liaison with other international standardizing bodies and user groups, and undertaking long-term planning. JTC 1 has been given the further task of developing common ISO-IEC procedures for standardization of information technology, based on the proposals of the ISO-IEC harmonization working group. International standards developed by JTC 1 are published under the ISO and IEC logos.

JTC 1 subcommittees are listed in Table 8.1 below. A full description of JTC 1 extracted from the *ISO Memento* is given at the end of this chapter in Appendix 8B (JTC 1 Operations and Subcommittees).

8.3 INTERNAL AND EXTERNAL LIAISONS

Internal Liaisons

JTC 1 has internal liaisons within ISO and IEC. The relevant liaison technical committees are shown in Table 8.2.

Liaison Between ISO-IEC and OSI User Groups

At the initiative of the ISO Executive Board, a special consultation on information technology systems was held at Geneva in October 1986 between Executive Board members and representatives of the *Open Systems Interconnection* (OSI) user groups.

The purpose of the consultation was to

- discuss how ISO could best assist the OSI user groups in achieving their international standardization objectives, and
- establish contacts to ensure that the international system will perform according to their expectations.

The meeting concluded that

Table 8.1. JTC 1 Subcommittees

AG	Advisory Group
SG	Functional Standardization
SC 1	Vocabulary
SC 2	Character Sets and Information Coding
SC 6	Telecommunications and Information Exchange Between Systems
SC 7	Software Development and System Documentation
SC 11	Flexible Magnetic Media for Digital Data Interchange
SC 14	Representations of Data Elements
SC 17	Identification and Credit Cards
SC 18	Text and Office Systems
SC 21	Information Retrieval, Transfer, and Management for Open Systems Interconnection (OSI)
SC 22	Languages
SC 23	Optical Digital Data Disks
SC 24	Computer Graphics
SC 25	Information Technology Equipment
SC 26	Microprocessor Systems
SC 27	Common Security Techniques
SC 28	Office Equipment

- a draft common statement of priorities for technical program areas would be circulated to participants;
- there was a need for a clear statement of ISO, IEC, and CCITT competence and unique expertise in the information technology field at the international level. The three organizations would specifically express their intention to work in a coordinated manner in responding to those parts of COS, MAP/TOP, POSI, SPAG, and SPAG Services programs that call for international standardization;
- a mechanism for executive level communications between the various parties present at the consultation would have to be devised, which would enable
    —the continuing definition of priorities, and
    —the definition of the role that is to be played by the various groups present in supporting the ISO-IEC work;

- a project for improvement of logistics and the speed of international standardization work in the field of information technology should be defined.

Priorities for Information Technology Standardization Work

A later consultation met at Geneva in March 1987, which also included representatives from the European MAP Users Group (EMUG) and from the newly formed European Group on Technical Office Protocols (OSITOP) together with the chairman and secretaries of the major subcommittees of ISO TC 97 and IEC.

The meeting decided that the basic information technology standards needed by the OSI user groups exist, or were being prepared, by ISO, IEC, and CCITT. Of twenty-two priority projects discussed, consulted groups expressed satisfaction with the target dates for sixteen projects and requested accelerated processing for six projects. The meeting also noted the need for further specification by the groups of their timing needs for particular standards to allow scheduling decisions to be taken by the appropriate ISO-IEC subcommittees.

Table 8.2. JTC 1 Liaison ISO and IEC Technical Committees

ISO		
TC		
	6	Paper, Board and Pulps
	37	Terminology (Principles and Co-ordination)
	39	Machine Tools
	46	Information and Documentation
	68	Banking and Related Financial Services
	154	Documents and Data Elements in Administration, Commerce and Industry
	159	Ergonomics
	184	Industrial Automation Systems
IEC		
TC		
	45	Nuclear Instrumentation
	65	Industrial—Process Measurement and Control
	74	Safety of Information Technology Equipment Including Electrical Business Equipment and Telecommunication Equipment

8.4 JOINT INFORMATION TECHNOLOGY RESOURCES SUPPORT GROUP

A proposal to set up a joint *Information Technology Resources Support Group* was endorsed by the IEC General Policy Committee (GPC) in April 1987, and by the ISO Executive Board in May 1987. The GPC's main function is to improve and facilitate the provision of administrative resources needed to service rapidly expanding work in the information technology field. The group, comprising executives of ISO member bodies and IEC national committees, having major resource support roles in information technology work, would provide the primary forum for executive-level dialogue with the major open systems interconnection user groups. The terms of reference of the support group are to

- maintain an overview of international information technology standardization work (including telecommunication), taking account of the expectations of vendors and users for both generic and applications standards,
- provide an ongoing forum for executive-level discussions with the main external organizations (including ITU, CCITT, and CCIR) concerned with information technology standardization at an international level,
- advise ISO and IEC councils, through the Executive Board and the General Policy Committee, on the provision and allocation of resources needed by ISO-IEC technical committees working on information technology standards.

The group's membership is subject to yearly review by the IEC General Policy Committee and the ISO Executive Board, and initially consists of

- a chairman who is the secretary-general of the ISO, or his or her appointed representative,
- a vice chairman who is the general secretary of the IEC, or his or her appointed representative, and
- members comprising the chief staff executives (the individual having ongoing responsibility for provision of national resources for administering international work) of the ISO member bodies and the IEC national committees holding the secretariats of technical committees and subcommittees involved in information technology.

8.5 INFORMATION TECHNOLOGY PROJECT-MONITORING DATABASE AND ELECTRONIC MAIL

A need has been seen for a unique and highly reliable database at the ISO's Central Secretariat, containing general information on stages of documents and production control, which would be accessible online to TC or SC secretariats in the information technology field and other interested parties. A further recommendation was made on the development of electronic transmission techniques (electronic mail) for selected Central Secretariat, TC, and SC communications.

A special working group set up by ISO TC 97 defined the content of the database with operational details in the form of advice for the ISO Central Secretariat. The additional annual costs were expected to be of the order of CHF 18,000 per secretariat, but appeared to be justified in view of the improvements the network would provide in managing the information technology standards programs.

8.6 ISO "FAST TRACK" PROCEDURES

The ISO has introduce a "fast track" procedure to facilitate the rapid adoption as ISO standards of effective *de facto* international standards prepared by other bodies, as outlined in Table 8.3 below.

If agreement to a text meeting the above requirements is impossible, the proposal has failed and the procedure is terminated. In either case, the WG shall prepare a full report which will be circulated by the secretariat of the relevant TC. The procedure has been used a number of times by JCT 1 to adopt standards prepared by other bodies.

8.7 JTC 1 DIRECTIVES AND TECHNICAL PROGRAM

Directives for the work of ISO-IEC JTC 1 were approved by ballot in October 1988. The index of the directives is in Appendix 8B, together with a summary of major provisions in the directives.

An excerpt from the JTC 1 Technical Program of 1989 is shown in Appendix 8C.

Table 8.3. Steps in the ISO's "Fast Track" Procedure

1	Any participating member and any Category A liaison organization of a concerned ISO TC may propose that an existing standard from any source be submitted directly for vote as an ISO DIS. The criteria for proposing an existing standard for the fast track procedure is a matter for each proposer to decide.
2	The proposal shall be received by the Central Secretariat, which will take the following actions.
2.1	To settle the copyright or trade mark situation with the proposer, so that the proposed text can be freely copied and distributed within ISO without restriction.
2.2	To assess, in consultation with the relevant secretariats, that ISO TC is competent for the subject covered by the proposed standard, and to ascertain that there is no evident contradiction with other ISO standards.
2.3	To distribute the text of the proposed standard as a DIS, indicating the TC to the domain of which the proposed standard belongs. In the case of particularly bulky documents, the Central Secretariat may demand the necessary number of copies from the proposer.
3	The period for combined DIS voting shall be six months. In order to be accepted, the DIS must be supported by 75% of the votes cast (abstention is not counted as a vote) and by an absolute majority of the participating members of the relevant TC.
4	At the end of the voting period, the comments received, whether editorial only or technical, will be considered by a working group appointed by the secretariat of the relevant TC.
5	If, after the deliberations of this WG, the requirements of step 3 above are met, the amended text will be sent to the Central Secretariat by the secretariat of the relevant TC for publication as an ISO standard.
6	If the proposed standard is accepted and published, its maintenance will be handled by the TC indicated according to step 2.3 above.

APPENDIX 8A

ISO-IEC JOINT TECHNICAL COMMITTEE JTC 1
OPERATIONS AND SUBCOMMITTEES
EXTRACTED FROM ISO MEMENTO, 1990

Conveners of working groups

A working group does not have a secretariat but an individual convener appointed by the parent committee to which he will report. In this list, and for practical reasons, the names of the conveners are replaced by the member body through which the convener can be reached. The following list also includes sub-committees which have not yet been allocated a secretariat, and working groups which do not yet have a convener.

Additional addresses for technical committee and sub-committee secretariats

The addresses of the technical committee and sub-committee secretariats are normally the same as those of the member bodies handling these secretariats (see addresses on pages 8 to 15). Where a TC or a SC secretariat has an additional address (see pages 134-142), the abbreviation of the member body is given in italics.

Chairmen of technical committees

When a technical committee nominates a permanent chairman, he need not be a national of the country holding the secretariat. Chairmen are normally appointed for a three-year term by the ISO Council and the figure in brackets shows the year at the end of which the term of office expires.

The chairman of the joint technical committee is appointed by both ISO and IEC Councils.

Responsables des groupes de travail

Un groupe de travail n'a pas de secrétariat mais un responsable désigné à titre individuel par le comité responsable auquel il fait rapport. Pour des raisons pratiques, les noms des responsables ont été remplacés dans la présente liste par ceux des comités membres auprès desquels ces responsables peuvent être atteints. La liste ci-après comprend également les sous-comités auxquels il n'a pas encore été attribué de secrétariat et les groupes de travail qui n'ont pas encore un responsable.

Adresses supplémentaires pour les secrétariats des comités techniques et des sous-comités

Les adresses des secrétariats des comités techniques et des sous-comités sont en général les mêmes que celles des comités membres qui en ont la charge (voir la liste des adresses aux pages 8 à 15). Si le secrétariat d'un TC ou d'un SC possède une adresse supplémentaire (voir pages 134-142), le sigle du comité membre est donné en italique.

Présidents des comités techniques

Lorsqu'un comité technique propose un président permanent, celui-ci ne doit pas nécessairement être un ressortissant du pays chargé du secrétariat. Le président est généralement nommé par le Conseil de l'ISO pour un mandat de trois ans et le chiffre figurant entre parenthèses indique l'année de l'échéance de son mandat.

Le président du comité technique mixte est nommé à la fois par les Conseils de l'ISO et de la CEI.

SWG	SIS	Conceptual model for electronic data interchange standards and services	Modèle conceptuel pour les normes et services d'échange de données informatisées
SG	**NNI**	**Functional standardization**	**Normalisation fonctionnelle**
SC 1	**AFNOR**	**Vocabulary**	**Vocabulaire**
WG 1	SCC	Advisory group for SC 1	Groupe consultatif du SC 1
WG 4	ANSI	Fundamental terms and office systems	Termes fondamentaux et bureautique
WG 5	DIN	Software	Logiciel
WG 6	SCC	Hardware, services and operations	Matériel, services et exploitation
WG 7	SCC	Communication	Communication
SC 2	**AFNOR**	**Character sets and information coding**	**Jeux de caractères et codage de l'information**

NOTE — Member body abbreviations in italics/Sigles des comités membres en italique, see/voir pages 135-143.

WG 1	SNV	Code extension techniques	Techniques d'extension de code
WG 2	ANSI	Multiple-octet coded character set	Jeu de caractères multi-octet
WG 3	SNV	7-bit and 8-bit codes	Codes à 7 et à 8 éléments
WG 6	—	Control functions	Fonctions de commande
WG 8	—	Coded representation of picture and audio information	Représentation codée de l'image et de l'information sonore
SC 6	**ANSI**	**Telecommunications and information exchange between systems**	**Téléinformatique**
WG 1	ANSI	Data link layer	Couche liaison de données
WG 2	BSI	Network layer	Couche réseau
WG 3	DIN	Physical layer	Couche physique
WG 4	AFNOR	Transport layer	Couche transport
WG 6	SNV	Private integrated services networking	Réseaux privés à intégration de services
SC 7	*SCC*	**Software development and system documentation**	**Élaboration du logiciel et documentation du système**
WG 1	ANSI	Symbols, charts and diagrams	Symboles, graphiques et diagrammes
WG 2	BSI	Software system documentation	Documentation du logiciel des systèmes
WG 3	SCC	Software engineering and software quality management	Génie du logiciel et gestion de la qualité du logiciel
WG 5	SCC	Reference model for software development	Modèle de référence pour le développement du logiciel
SC 11	**ANSI**	**Flexible magnetic media for digital data interchange**	**Support magnétique flexible pour l'échange de données numériques**
WG 3	DIN	Lower-level interface functional requirements and lower-level interfaces	Spécifications fonctionnelles d'interfaces à niveau inférieur et interfaces à niveau inférieur
SC 14	**SIS**	**Representations of data elements**	**Représentations des éléments de données**
WG 1	SIS	Standardization guidelines for the representation of data elements	Principes directeurs de normalisation pour la représentation des éléments de données
WG 3	SIS	Terminology	Terminologie
WG 4	ANSI	Co-ordination of data element standardization	Coordination de la normalisation des éléments de données
SC 17	*BSI*	**Identification cards and related devices**	**Cartes d'identification et dispositifs associés**
WG 1	DIN	Physical characteristics and test methods for ID-cards	Caractéristiques physiques et méthodes d'essais des cartes d'identification
WG 3	ANSI	Passport cards	Cartes passeport
WG 4	AFNOR	Integrated circuit card	Carte avec circuit intégré
WG 5	ANSI	Registration Management Group (RMG)	Comité de gestion pour l'enregistrement (RMG)
WG 8	BSI	Integrated circuit cards without contacts	Cartes avec circuit intégré sans contact
SC 18	**ANSI**	**Text and office systems**	**Bureautique**
WG 1		User requirements and SC 18 management support	Besoins des utilisateurs et soutien à la gestion du SC 18
WG 3	BSI	Document architecture	Architecture des documents
WG 4	AFNOR	Procedures for text interchange	Procédures pour l'échange de textes
WG 5	SCC	Content architectures	Architectures du contenu
WG 8	ANSI	Text description and processing languages	Description du texte et langages de traitement
WG 9	ANSI	User/systems interfaces and symbols	Interfaces et symboles utilisateur/systèmes
SC 21	**ANSI**	**Information retrieval, transfer and management for open systems interconnection (OSI)**	**Accès, transfert et gestion de l'information pour l'interconnexion des systèmes ouverts (OSI)**
WG 1	AFNOR	OSI architecture	Architecture OSI
WG 3	SCC	Database	Base de données
WG 4	JISC	OSI management	Administration OSI
WG 5	BSI	Specific application services	Services d'applications spécifiques
WG 6	ANSI	OSI session, presentation and common application services	Session, présentation et services commun d'application OSI
WG 7	NNI	Basic reference model of open distributed processing	Modèle de référence de base pour traitement distribué ouvert
SC 22	*SCC*	**Languages**	**Langages**
WG 2	BSI	Pascal	Pascal
WG 3	SCC	APL	APL
WG 4	ANSI	Cobol	Cobol
WG 5	ANSI	Fortran	Fortran
WG 8	ANSI	Basic	Basic
WG 9	ANSI	Ada	Ada
WG 11	ANSI	Binding techniques	Techniques d'association
WG 13	BSI	Modula 2	Modula 2
WG 14	ANSI	C	C
WG 15	ANSI	POSIX	POSIX

NOTE — Member body abbreviations in italics/*Sigles des comités membres en italique*, see/*voir* pages 135-143.

WG 16	AFNOR	LISP	LISP
WG 17	BSI	Prolog	Prolog
WG 18	ANSI	Forms interface management system (FIMS)	Gestion de système forme interface
SC 23	*JISC*	**Optical disk cartridges for information interchange**	**Cartouches de disques optiques pour échange d'information**
WG 1		Permanent editing committee	Comité de rédaction permanent
WG 2	ANSI	90 mm and 130 mm rewritable ODCs	ODC de 90 mm et 130 mm réutilisables
WG 3	AFNOR	300 mm WORM ODCs	ODC-WORM de 300 mm
WG 4	NNI	130 mm WORM ODCs	ODC-WORM de 130 mm
SC 24	**DIN**	**Computer graphics**	**Infographie**
WG 1	ANSI	Architecture	Architecture
WG 2	SNV	Application programming interface	Interface de programmation selon application
WG 3	BSI	Metafiles and device interface	Métafichier et interface du dispositif
WG 4	ANSI	Language bindings	Interfaces langages
WG 5	DIN	Validation, testing and registration	Validation, essais et enregistrement
SC 25	**DIN**	**Interconnection of information technology equipment***	**Interconnexion des techniques relatives à l'information***
WG 1		Home electronic systems	Systèmes électroniques domestiques
WG 2		Fibre optic connections for information technology equipment	Connexions par fibre optique des appareils de traitement de l'information
WG 3		Customer premises cabling	Câblage des locaux d'utilisateurs
WG 4		Interconnection of computer systems and attached equipment	Interconnexion des systèmes informatiques et des appareils raccordés à ces systèmes
SC 26	*JISC*	**Microprocessor systems**	**Systèmes à microprocesseurs**
WG 1		Definitions of microprocessor instructions and their mnemonic representation	Définitions des instructions pour microprocesseurs et de leur représentations mnémoniques
WG 4		Architecture	Architecture
WG 6		Revision of Publication 821	Révision de la Publication 821
WG 7		Microprocessor systems quality assessment	Evaluation de la qualité des systèmes à microprocesseurs
SC 27	**DIN**	**Security techniques**	**Techniques de sécurité**
WG 1	BSI	Secret key algorithms and applications	Algorithmes à clés secrètes et applications
WG 2	AFNOR	Public key cryptosystem and mode of use	Système de chiffrement des clés révélées et mode d'utilisation
SC 28	**SNV**	**Office equipment**	**Équipements de bureau**

* Provisional / provisoire

NOTE — Member body abbreviations in italics / Sigles des comités membres en italique, see / voir pages 135-143.

APPENDIX 8B

DIRECTIVES FOR THE WORK OF ISO-IEC JOINT TECHNICAL COMMITTEE 1 (JTC 1) ON INFORMATION TECHNOLOGY

INTRODUCTION

On 20 June 1988 the Directives for the work of ISO-IEC Joint Technical Committee 1 (JCT 1) on Information Technology were issued for ballot. The purpose of that document was to provide, under a single cover, a complete set of procedures covering the development, adoption and publication of common ISO-IEC International Standards developed by JCT 1. It was designed to be used by all participants in JCT 1, including its subcommittees and working groups.

The Directives were produced by a standing committee known as the Special Working Group on Procedures (SWG-P). The procedures contained in the Directives were formally adopted at the Plenary Session of JCT 1 when it met in Tokyo in November 1987. Following resolution H taken at the Advisory Group meeting in Washington, DC, the Directives were submitted for letter ballot of JTC 1 national bodies who had to complete and return the letter ballot to the Secretariat by 20 October 1988.

ORGANIZATIONAL STRUCTURE

Secretariat and Chairman

The Secretariat is appointed from among the P-members of the JCT 1. The Chairman is nominated by the JCT 1 Secretariat endorsed by JCT 1 and formally appointed by the ISO and IEC Councils. A Vice Chairman is appointed for each of the four technical groupings of subcommittees.

Membership

Membership of JCT 1 comprises National Bodies which are countries eligible for membership through their existing membership as member bodies of ISO or as national committees of IEC, which participate actively as P-members or are kept informed of progress of the work as O-members.

Subcommittees (SCs)

JCT 1 establishes SCs charged with the study of particular parts of its programme of work.

Subcommittees comprise at least five participating members and are numbered from JCT–1 SC 1.

The scope for each SC is a concise statement precisely defining the limits of the work of the SC within the scope of JCT 1.

The Secretariat of an SC is appointed by JCT 1 from among the P- members of that SC. Chairmen are nominated by the secretariat, endorsed by the SC, and appointed by JCT 1 for nominally three years.

The SCs are structured, subject to modification by JCT 1, into the following technical groupings, to ease the co-ordination of work.

GROUPINGS OF JCT 1 SC

Group 1 - Application elements (AE)
- SC 1   Vocabulary
- SC 7   Software development and systems documentation
- SC 14   Representation of data elements
- SC 22   Languages
- SC 28   Office equipment

Group 2 - Equipment and media (EM)
- SC 11   Flexible magnetic media for digital data exchange
- SC 17   Identification and credit cards
- SC 23   Optical digital data disks

Group 3 - Systems support (SS)
- SC 2   Character sets and information coding
- SC 24   Computer graphics
- SC 25   Information technology equipment
- SC 27   Common security techniques

Group 4 - Systems (S)
- SC6   Telecommunications and information exchange between systems
- SC18   Text and office systems

SC 21 Information retrieval, transfer and management for open systems interconnection (OSI)

SC 26 Microprocessor systems

Working Groups (WGs)

JCT 1 or its SCs may establish a working group only if there is one or more approved projects for which it is responsible.

Membership is open to all P-members of the parent body and to organizations in liaison category A. Members of the WG are appointed by their National Bodies or A-liaison organizations.

WG conveners are selected and appointed by the subcommittee, the convener being endorsed by his National Body on the basis that the convener has the necessary resources and administrative support to carry out the responsibilities assigned. All WG convenerships are for three year terms, but there may be a reappointment and re-endorsement by the parent body.

The SC Secretariat notifies the JCT 1 Secretariat and the Information Technology Task Force (ITTF) of the names and addresses of appointed conveners.

The SC may assign responsibility for the administration of a WG either to a convener or to a secretariat. The secretariat of a WG may be either a National Body or an organization residing within the same country as the convener.

The ITTF is kept informed of arrangements and advised of address details of the person responsible for the administration of the WG.

SCs periodically review the performance of their WGs by considering

- whether the work items are progressing in accordance with established target dates
- whether the experts nominated by the National Bodies which agreed to participate in the development of the work items continue to participate in the work by attendance at meetings and/or submission of contributions.

WG members are to indicate whether the views expressed reflect National Body positions or personal opinions.

JCT 1 and its SCs may establish other working groups (*known as special working groups*) to undertake specific tasks, generally between meetings, the tasks being defined at the meeting of the parent body.

These groups include an *ad hoc* group, a rapporteur group, a drafting group and an editing group.

In special cases within JCT 1 a *joint working group* may be established to undertake a specific task in which more than one SC is strongly interested.

INFORMATION TECHNOLOGY TASK FORCE (ITTF)

The ITTF is a joint group provided by the IEC Central Office and the ISO Central Secretariat to provide joint support for the activities of JCT 1, but the ITTF is not part of the JCT 1 structure.

The ITTF is responsible for the day to day planning and co-ordination of the technical work of JCT 1 relative to IEC and ISO, and supervises the application of the ISO and IEC constitutions and rules of procedure. The ITTF satisfies itself that particular investigations are followed up and that the time limits are complied with. The ITTF must always be fully informed regarding technical work envisaged, in progress and completed.

The ITTF advises the secretariat and the secretary on any point of procedure and assists in the technical co-ordination and harmonization of work. In relation to international standards the ITTF has the responsibility of

- registration of CDs
- checking and editing of DIS before their submission to National Bodies
- reproduction and circulation of DIS to National Bodies for approval and to organizations in liaison for information and comment
- administering the voting of National Bodies on DIS
- communicating the voting results and related comments to the JCT 1 Secretariat
- acceptance on behalf of the ISO Secretary-General and IEC General Secretary of final tests for publication as International Standards
- ensuring the correct presentation of international standards and printing, distribution and sale of international standards.

The ITTF is also responsible for

- maintaining up to date records showing the participant categories (P and O) of National Bodies in JCT 1 and each SC

- maintaining-up-to date records of liaisons established for JCT 1 and its subcommittees
- co-ordinating the meetings of JCT 1 and SCs relative to other ISO and IEC TCs
- convening meetings of JCT 1 and SCs
- dealing with questions concerning relations with external organizations.

JCT 1 STANDING ORGANISATIONS

Advisory Group (AG)

The purpose of the AG, established at the first JCT 1 plenary meeting, is to assist the JCT 1 Chairman and Secretariat in the co-ordination of subcommittee activities, in the preparation and monitoring of the overall programme of work and meeting schedules, and in the discussion and resolution of inter-committee problems and issues.

There are a number of issues of concern for AG meetings listed in the directives, for example liaisons between SCs, identifying new areas of work, developing procedures addressing common SC non-technical issues such as procedures for maintenance of standards requiring rapid amendment. The AG is responsible for preparing advice or guidance, or proposals for consideration by the JCT 1 Chairman and Secretariat and by National Bodies.

Membership of the AG is open to all P-members of JCT 1. Meetings of the AG are convened by the Secretariats as frequently as needed to expedite the operation of JCT 1.

Special Group on Functional Standardization (SG-FS)

SG-FS deals with functional standards submitted by organizations which develop such standards and wish to have them adopted as International Standard Profiles (ISP). An ISP is an internationally agreed standard, together with options and parameters, necessary to accomplish a function or a set of functions.

Membership of SG-FS is open to all P-members of JCT 1, which functions as an SC and uses special procedures for ISP development accepted by the IEC and ISO Councils. In all matters SG-FS has the same relationship with the AG as all SCs, but is not included in any grouping.

Special Working Group on Strategic Planning (SWG-SP)

JCT 1 has established a special working group on strategic planning which operates under the auspices of the AG and is open to all P-members of JCT 1. There is a planning function within each subcommittee of JCT 1 to maintain a plan for the future direction of its programme within its area of work.

Special Working Group on Procedures (SWG-P)

The SWG-P operates under the auspices of the AG. The SWG-P co-operates with ISO Central Secretariat and IEC Central Office on the development of detailed harmonized procedures, and reviews the procedural requirements of JCT 1. It also monitors the implementation of existing procedures, evaluating their effectiveness and making proposals for changes where found necessary. Membership is open to all P-members of JCT 1.

Special Working Group on Registration Authorities (SWG-RA)

SWG-RA operates under the auspices of the AG and membership is open to all P-members of JCT 1.

PROGRAMME OF WORK

JCT 1 has established a programme of work considering requests for international standards initiated by sources within and outside JCT 1. The programme of work, within the scope agreed by the Councils, consists of a detailed list of all work items for study. The selection of items is subject to close scrutiny in accordance with policy objectives and resources of IEC-ISO and is governed by economic, social and technical consideration. Each item in the programme of work is separately approved by letter ballot of P-members of JCT 1. This requirement does not however prevent initiation of discussion of technical documents pertaining to a proposed new item pending approval of the item by letter ballot of P-members.

New Work Item (NWI)

All proposals for NWIs are to be accompanied by projected target dates for completion and in voting on any NWI proposal National Bodies are required to answer

- do you accept the definition of the NWI
- do you support the NWI
- do you commit yourself to participate in the development of NWI by sending representatives to meetings regularly and submitting written contributions

- are you able to commit resources to this NWI without detracting from your contributions to the existing work of JCT 1
- do you have a major contribution or a reference document on this topic ready for submission to an SC
- will you have such a contribution within 90 days
- are you able to offer a project editor who will dedicate his efforts to the advancement and maintenance of this project.

Each item in the programme of work is to be given a serial number which is retained in the programme of work under that serial number until the work on that item is completed or its deletion has been agreed.

If a work item remains at stage 1 three years after its inclusion in the programme of work, it is cancelled automatically unless retention is confirmed by a majority of P-members voting before the end of the period. Similarly, an item at stage two on which no apparent progress has been made after five years may be deleted unless its retention has been confirmed.

Formal Description Techniques (FDTs)

There have often been difficulties in the definition of standards because of ambiguities in the English language. As a result of this, the directives of JCT 1 advise using FDTs. FDTs are used to describe complex functions in relatively precise, formal and concise language so that no ambiguities exist. Only one meaning of a statement is then possible.

If experience and resources for FDs is lacking in National Bodies, the development of standards must be based on conventional natural language approaches leading to standards where the natural language description is the definitive standard. SCs are encouraged to develop FDs of their standards as these efforts may contribute to the quality of the standards by detecting defects, may provide additional understanding to readers and should support the evolutionary introduction of FDTs. An FD produced by an SC that can be considered to represent faithfully a significant part of the standard or the complete standard should be published with the standard. JCT 1 considers that the development of standards should be accompanied and supported by the development of FDs with the object of improving and supporting the structure, consistency and correctness of the natural language description. An assurance must exist that the application of FDTs does not unnecessarily restrict the freedom of implementation.

Where FDs become part of the standards together with the natural language description, the SC should indicate in the standard which description should be treated as the definitive version. Where discrepancy exists between a natural language description and an FD the discrepancy should be resolved by changing or improving the natural language description.

Development and Acceptance of FDTs

Only standard FDTs or FDTs in the process of being standardized should be used in formal descriptions (FDs) of standards. If an FD is to be developed for a new standard, the FD should be progressed according to the same timetable as the rest of the standard.

All proposals for standardizing new FDTs are subject to the NWI voting procedure and

- the need for the FDT must be demonstrated
- it must be shown that it is based on a significantly different model from that of any existing FDT, and
- the usefulness and capabilities of the FDT must be demonstrated.

Target Dates and Priorities

JCT 1 establishes target dates for the registration of the first draft proposal on each item of the programme of work and for the transmission from draft proposal to draft International Standard.

Target dates are recorded at the ITTF and kept under periodic review by JCT 1 and are amended as necessary.

Each SC must establish priorities and establish timetables and target dates for all work items assigned to that SC. In particular there must be target dates for

- registration of the first CD (and subsequent CDs)
- submission of texts for DIS processing, and
- publication of International Standards.

Progress Control

JCT 1 is to ensure that the planned programme is pursued and that, as far as possible, established target dates are met. Control is to be exercised over each separate work item for each stage in the procedure.

Each subcommittee reviews its progress against the target dates at regular intervals and amends target dates where necessary, justification being provided to JCT 1 for such amendments.

A written report on progress of work, target dates, progress reports and the deletion or redefinition of work items is submitted annually by each SC.

Annual Report

The JCT 1 Secretariat prepares an annual report describing the work of JCT 1, its SCs and WGs during the year ending 31 December.

The contents of the annual report is prescribed and includes

- a description of the work of the SCs (i.e., the SC's scopes) and a description of the WGs (i.e., the WG's terms of reference)
- the latest version of the committee's programme of work together with information as to the development stage of various work items
- target dates for appropriate stages of development (e.g., registration of the first CD, subsequent CDs and submission of text for DIS processing).

Part 2 of the annual report is a management report which

- evaluates the performance of each SC against its target dates
- delineates the problem areas
- identifies the action being taken to address the problems, and
- evaluates the effectiveness of liaisons.

There is substantial responsibility on the JCT 1 Secretariat and each SC secretariat. Each secretariat must act in all respects as the international secretariat and not be influenced by national considerations in the pursuit of work. The National Body, in maintaining strict neutrality, is to distinguish sharply between proposals it makes as a National Body and proposals made in its capacity as secretariat.

Members of JCT 1 and its Subcommittees

P-members of JCT 1 and its SCs have an obligation to take an active part in the work of the JCT 1 or SC, and to vote within the time limits laid down on all questions and to attend meetings.

If a P-member is persistently inactive and has failed to make a contribution to two consecutive Plenary meetings of JCT 1 or the SC concerned, or has failed to vote on questions submitted for voting, the Secretaries-General are to remind the National Body of its obligation to take an active part in the work of the committee. In the absence of a positive response to this reminder, the member shall automatically have its status changed to that of O-member.

PREPARATION AND ADOPTION OF INTERNATIONAL STANDARDS

The social and economic long-term benefits of an international standard should justify the total cost of preparing, adopting and maintaining the standard. Technical consideration should demonstrate that the proposed standard is technically feasible and timely and that it is not likely to be made obsolete quickly by advancing technology or to inhibit the benefits to users of technological advance.

With consensus in mind, a draft for an international standard must achieve substantial support in JCT 1 before being further processed.

Terminology

Terms that are used for successive documents drawn up on a single subject are:

Proposal for a New Work Item (NWI) — a proposal circulated by the Secretariat of JCT 1 to the P-members with a view to the inclusion of the work item at stage 2 in the programme of work.

Working Draft — a document pertaining to a work item at stage 2 circulated by the secretariat of JCT 1 or any of its SCs or by the convener of a WG to its members, with a view to the subsequent presentation of a Committee Draft.

Committee Draft (CD) — a proposal for an International Standard registered at the ITTF to which a CD number has been allocated and submitted to members of JCT 1 or one of its SCs for vote and comment. Successive Committee drafts on the same subject are marked "first committee draft," "second committee draft," etc.

Draft International Standard (DIS) — a draft proposal which has received substantial support from the P-members of a TC for publication as a Draft International Standard and has been registered at the ITTF for circulation to National Bodies for approval.

International Standard — a DIS which has been adopted by the majority of P-members of JCT 1 and has been approved by at least 75% of the National Bodies voting and accepted for publication by the JCT 1 Secretariat and ITTF.

Stages of Technical Work

The successive stages of the technical work are

Stage 1 (proposal stage):	A NWI proposal is under consideration
Stage 2 (preparatory stage):	A working draft is under consideration
Stage 3 (committee stage):	A CD is under consideration
Stage 4 (approval stage):	A DIS is under consideration
Stage 5 (publication stage):	An International Standard is being prepared for publication.

Specific steps are taken to avoid delay. JCT 1 avoids discussion of a document successively at more than two levels of the three levels — WG/SC/TC. The two levels are the expert level where technical proposals are discussed and drafts prepared (i.e., WG or SC) and the committee level (i.e., SC or JCT 1) at which final National Body vote on the draft is expressed within JCT 1.

Preparation of Committee Draft (CD) and Draft International Standards (DIS)

Preparation of Committee Draft (CD)

A working draft has reached the stage of Committee Draft (CD) when

- the main elements have been included in a document
- it is presented in a form which is essentially that envisaged for the future International Standard
- it has been dealt with at least once by JCT 1 or by a working body of JCT 1
- JCT 1 or one of its SCs has decided in a resolution during a meeting or by correspondence that the working draft be forwarded to the ITTF for registration as a CD.

Once the ITTF registers it as a CD and allocates a serial number to it, the work has reached stage 3. The number remains the same throughout the reporting stages and for the published International Standard.

The JCT 1 or SC secretariat circulates the Committee Draft. The introductory note indicates the sources used as a basis for the proposal, the background and aim of the proposal, an outline of the technical justification of the Committee Draft and whenever appropriate, a summary of the technological data on which it is based. Liaisons with other interested TCs, SCs or organizations are also stated.

Finalization of Committee Draft (CD)

P-members and TCs and organizations in liaisons are asked to submit their comments by a specified date. In the case of CDs the date should be approximately three months from the date of circulation. Comments and votes are sent to the secretariat of JCT 1 or SC within the period specified, and summarized by the secretariat and distributed. The secretariat should also distribute a report clearly indicating the action taken as a result of the comments received and should circulate, if necessary, a further CD.

If a CD is considered at a meeting but agreement on it is not reached on that occasion, the secretariat distributes a revised CD. Consideration of successive CDs continues until the substantial support of the P-members of the committee has been obtained or a decision to abandon or defer the project has been reached.

The JCT 1 or SC secretariat has the responsibility of judging when substantial support has been obtained. Attention is given not only to numerical voting results, but also to success in resolving negative votes.

If substantial support for a CD is obtained at a meeting or by correspondence as it was distributed or subject to necessary corrections, the CD after modification is submitted directly to the ITTF.

When substantial support has been obtained in JCT 1, the secretariat of JCT 1, or SC if authorized by JCT 1, submits the CD, normally within a maximum of three months, to the ITTF for registration as a DIS and circulation to National Bodies for approval.

Whenever appropriate SCs entrusting tasks to WGs or SWGs should empower them to produce on behalf of the SCs the CD or DIS text for direct submission to ITTF via the SC secretariat.

If the substantial support of the P-members is obtained in a SC, the CD shall be submitted by the SC secretariat within a maximum of three months to the ITTF for registration and circulation as a DIS.

Whenever possible DIS are submitted simultaneously to voting by the P-members of JCT 1 and by the National Bodies (combined voting procedure).

Draft International Standards (DIS)

When the substantial support of the P-members of the SC has been obtained, the SC secretariat sends within three months to the ITTF

- the final text of the CD for circulation as a DIS
- originals or clear prints of any figures or graphs, and
- an explanatory report.

The explanatory report contains

- a brief history of the draft
- a report on how substantial support of the P-members of the SC was obtained or in cases where a formal vote of the P-members of JCT 1 has taken place, a record of the voting on the CD listing those P-members of the SC who voted in favor, those who voted against and those who did not vote
- a brief statement of all technical objections which have not been resolved and the reasons why it has not been possible to resolve them
- in the case of a revision of an existing International Standard, identification of clauses in the previous edition of the International Standard now proposed for technical revision.

The ITTF registers the CD as a DIS and circulates the draft received together with the explanatory report to all National Bodies for approval within six months, if necessary implementing the combined voting procedure.

National Bodies reply by

- approval of the technical content of the DIS as presented (with editoral or other comments appended)
- disapproval of the DIS for reasons stated, or
- abstention.

In the case of disapproval, acceptance of objections will change the vote to approval after reference to an agreement of the National Body concerned.

If the DIS is not adopted by the majority of P-members and/or approved by 75% of the National Bodies voting, it cannot go forward and the matter is referred back to the SC secretariat. A new draft may be prepared for submission to National Bodies provided that the time for voting is reduced to three months. The same procedure may be repeated until a DIS has been adopted by a majority of the P-members and approved by 75% of the National Bodies voting.

In the absence of the necessary majority, JCT 1 may decide at any stage to request the publication of the draft as a technical report, if the majority of the P-members agree.

Acceptance of International Standards for Publication

When a DIS has been adopted by the majority of the P-members of JCT 1 and has been approved by 75% of the National Bodies voting, the ITTF informs the JCT 1 and SC secretariat accordingly and communicates to the SC secretariat the results of voting and the comments made by National Bodies.

Within a maximum of four months or such lesser period as is agreed with the ITTF the SC secretariat or designated project editor prepares

- a revised text of the DIS, and
- a report indicating the action taken on the technical and other major comments made by National Bodies and if any objections have not been resolved, a clear statement of the reasons.

The revised text of the DIS and the report is sent to the ITTF with the decision of the SC secretariat to publish the international standard.

The SC secretariat informs the ITTF if the secretariat cannot meet the agreed deadline but if the revised text is not returned within a period of twelve months the DIS is resubmitted for National Body voting.

The ITTF prepares a final report giving the results of National Body voting and fully reflecting the comments made by National Bodies and the action taken thereon by the SC secretariat. The final report is circulated by the ITTF to all P-members of JCT 1 and all National Bodies are informed that the DIS has been accepted for publication. If no National Body requests within six weeks further consideration of its comments, the ITTF shall proceed with publication of the International Standard.

Periodic Review

On request by a National Body or the Secretaries-General and in any case at not more than five-yearly intervals, JCT 1 shall review the International Standards for which it is responsible with a view to deciding (by a majority of the P-members voting in a meeting or by correspondence) whether they should be

- confirmed
- revised, or
- withdrawn.

The review is to include an assessment of the degree to which the standards have been applied in practice.

Revision

The ITTF is informed as soon as a revision or amendment of an International Standard is envisaged by JCT 1. Steps for a revision start with stage 2. JCT 1, however, or one of its SCs by a vote of its P-members or at a meeting, may decide that a proposed revision or amendment is of a relatively minor importance and the revised International Standard or agreed amendments may be submitted directly to the ITTF for publication. Where continuous updating of an International Standard is required, JCT 1 may request the establishment of a maintenance agency. Procedures for rapid amendment of JCT 1 standards are to be applied to those standards for which proper implementation is dependent on the careful but rapid promulgation of errata or amendments as faults are detected.

There is also provision for circulation and consideration of defect reports.

FAST TRACK PROCEDURE

Proposal and Voting

Any P-member or any Category A liaison organization of JCT 1 may propose that an existing standard from any source be submitted directly for vote as a DIS. The criteria for proposing an existing standard for fast-track procedure are for each proposer to decide.

The proposal is received by the ITTF which will then

- settle the copyright and/or trade mark situation with the proposer so that the proposed text can be freely copied and distributed without restriction
- assess in consultation with JCT 1 Secretariat that JCT 1 is the competent committee for the subject covered in the proposed standard and ensure that there is no evident contradiction with other ISO/IEC standards
- distribute the text of the proposed standard as a DIS.

The period for combined DIS voting shall be six months and the DIS must be supported by 75% of the votes cast and by the majority of P- members of JCT 1 for acceptance. At the end of the voting period the comments received whether editorial only or technical, are dealt with by a WG appointed by JCT 1 Secretariat. If after the deliberations of this WG, the voting requirements are met, the amended text is sent to ITTF by the JCT 1 Secretariat for publication as an International Standard. If it is impossible to agree to a text meeting these requirements, the proposal has failed and the procedure is terminated. In either case the WG prepares a full report which is circulated by the JCT 1 Secretariat.

Handling of Comments on Fast-Track Processed Draft International Standards

The ITTF notifies the JCT 1 Secretariat that a DIS has been registered for fast-track processing. The JCT 1 Secretariat then identifies the JCT 1 SC that should deal with the DIS, and informs the SC secretariat of the fast-track processed DIS number, title, and ballot period dates, and sends the SC secretariat a copy of the DIS.

The SC secretariat informs the SC National Bodies and makes plans for the handling of ballot results through the formation of a WG by

- scheduling a WG meeting to consider any comments on the DIS
- appointing a convenor for the WG meeting
- appointing a project editor for the DIS, and
- notifying the SC National Bodies of the WG meeting dates, location, convener and project editor.

The SC secretariat circulates to the SC National Bodies the DIS ballot results and any comments. At the WG meeting decisions should be reached preferably by consensus. If a vote is unavoidable the vote of the National Bodies is taken in accordance with normal JCT 1 procedures. The convenor prepares a final report in co-ordination with the project editor who prepares the final DIS text in the case of acceptance.

The time period for these steps is

- a total of two months for the ITTF to send the results of the vote to the JCT 1 Secretariat and to the SC secretariat and for the latter to distribute it to its National Bodies
- not less than ten weeks prior to the date of the WG meeting for distribution of the voting results and any comments
- not later than one month from the WG meeting for distribution by the SC secretariat of the final report and the final DIS text in case of acceptance.

PUBLICATION OF TECHNICAL REPORTS

The primary duty of JCT 1 is the preparation and review of International Standards. When, despite repeated efforts within JCT 1, the substantial support or necessary majority cannot be obtained for submission of a CD for registration as a DIS or for approval of a DIS at National Body voting stage, JCT 1 may decide to request publication of the document in a form of a technical report. The reasons why the required support could not be obtained should be mentioned in the document (type 1).

When the subject in question is still under a technical development or where for any other reason there is a possibility of an agreement at some time in the future, JCT 1 may decide that the publication of a technical report would be more appropriate (type 2).

When JCT 1 has collected data of a different kind from that which is normally published as an international standard, for example, factual information obtained from a survey carried out among the National Bodies, JCT 1 may propose to the Secretaries-General that the information be published as a technical report (type 3). Technical reports of types 1 and 2 contain

- historical background
- explanation of reasons why JCT 1 considered it necessary to publish a technical report instead of an International Standard, and

- technical content.

When the majority of P-members of JCT 1 have agreed to the publication of a technical report, it is submitted by the JCT 1 secretariat to the ITTF normally within two months but is subject to review by JCT 1 not later than three years after its publication.

APPENDIX 8C

JTC 1 TECHNICAL PROGRAM, 1989

REFERENCE	T A ED	TITLE	TITRE	AC PR
				STAGE
JTC 1		**Information technology**	**Technologies de l'information**	
		UDC 681.3	CDU 681.3	
DP 646	92-06 3	Information processing -- ISO 7-bit coded character set for information interchange (Revision of ISO 646:1983)	Traitement de l'information -- Jeu ISO de caractères codes à 7 éléments pour l'échange d'information (Revision de l'ISO 646:1983)	2.1 1.1 89-05
DP 821	91-06 2	Microprocessor system bus for 1 to 4 byte data (Revision of IEC 821:1987)	Bus système à microprocesseurs pour données de 1 a 4 octets (Revision de la CEI 821:1987)	2.1 89-06
DP 1539	91-10 2	Information processing systems -- Programming languages -- FORTRAN (Revision of ISO 1539:1980)	Systèmes de traitement de l'information -- Langages de programmation -- FORTRAN (Revision de l'ISO 1539:1980)	2.1 1.1 87-10
DIS 1863	89-03 2	Information processing -- 9-track, 12,7 mm (0.5 in) wide magnetic tape for information interchange using NRZ 1 at 32 ftpmm (800 ftpi), 32 cpmm (800 cpi) (Revision of ISO 1863:1976)	Ce DIS est distribué en version anglaise seulement	4.4 4.1 89-03
DIS 2110	89-10	Information technology -- Data communication -- 25-pole DTE/DCE interface connector and contact number assignments (Revision of ISO 2110:1980)	Communication de données -- Description du connecteur 25 broches à la jonction entre ETTD et ETCD et affectation des broches (Revision de l'ISO 2110:1980)	5.1 4.4 89-06
DP 2382-1	92-06 3	Information processing systems -- Vocabulary -- Part 01 : Fundamental terms (Revision of ISO 2382-1:1984)	Systèmes de traitement de l'information -- Vocabulaire -- Partie 01 : Termes fondamentaux (Revision de l'ISO 2382-1:1984)	2.1 1.5 89-01
DIS 2382-7	90-02 2	Information processing systems -- Vocabulary -- Part 07 : Computer programming (Revision of ISO 2382-7:1977) Bilingual edition	Systèmes de traitement de l'information -- Vocabulaire -- Partie 07 : Programmation des ordinateurs (Revision de l'ISO 2382-7:1977) Édition bilingue	4.2 4.3 89-02
DP 2382-16	91-06 2	Information processing systems -- Vocabulary -- Part 16: Information theory (Revision of ISO 2382-16:1978)	Systèmes de traitement de l'information -- Vocabulaire -- Partie 16: Theorie de l'information (Revision de l'ISO 2382-16:1978)	2.8 2.1 89-01
DIS 2382-20	90-06	Information technology -- Vocabulary -- Part 20 : System development Bilingual edition	Technologies de l'information -- Vocabulaire -- Partie 20 : Developpement de systeme Édition bilingue	4.2 4.1 89-05
DP 2382-23	92-06	Information processing systems -- Vocabulary -- Part 23: Text processing	Systèmes de traitement de l'information -- Vocabulaire -- Partie 23: Traitement de texte	2.1 1.5 89-03
DP 2382-25	91-02	Information processing system -- Vocabulary -- Part 25: Local area networks	Systèmes de traitement de l'information -- Vocabulaire -- Partie 25: Reseaux locaux	2.3 2.2 89-05
DIS 2593	91-06 3	Information technology -- Data communication -- 34-pole DTE/DCE interface connector and contact number assignments	Technologie de l'information -- Communication de données -- Connecteur d'interface ETTD/ETCD à 34 pôles et affectation des numeros de contact (Revision de l'ISO 2593:1984) (DIS distribué en version anglaise seulement)	3.5 3.1 89-05
DAD 3309	90-11	Information processing systems -- Data communication -- High-level data link control procedures -- Frame structure ADDENDUM 1: Start/stop transmission	Systèmes de traitement de l'information -- Communication de données -- Procedures de commande de liaison de données à haut niveau -- Structure de base ADDITIF 1: Transmission commencer/arrêter (DAD distribué en version anglaise seulement)	3.5 3.1 89-03
DIS 3788	89-03 2	Information processing -- 9-track, 12,7 mm (0.5 in) wide magnetic tape for information interchange using phase encoding at 126 ftpmm (3200 ftpi), 63 cpmm (1600 cpi) (Revision of ISO 3788:1976)	Ce DIS est soumis en version anglaise seulement	4.4 4.1 89-03

REFERENCE	T A ED T I T L E	T I T R E	STAGE AC PR

JTC 1 (CONTINUED)/(SUITE)

DAD 4335	88-04	Information processing systems -- Data communication -- High-level data link control procedures -- Consolidation of elements of procedures ADDENDUM 2 : Enhancement of the XID function utility (Will be incorporated in a new edition of ISO 4335)	Systèmes de traitement de l'information -- Communication de données -- Procédures de commande de liaison de données à haut niveau -- Consolidation des éléments de procédures ADDITIF 2 : Amélioration du rôle de la fonction XID (Sera incorporé à une nouvelle édition de l'ISO 4335)	4.4 4.2 88-06
DAD 4335	90-11	Information processing systems -- Data communication -- High-level data link control elements of procedures ADDENDUM 3: Start/stop transmission	Systèmes de traitement de l'information -- Communication de données -- Procédures de commande de liaison de données à haut niveau ADDITIF 3: Transmission commencer/arrêter (DAD distribué en version anglaise seulement)	3.5 3.1 89-03
DP 4873	92-06 3	Information processing -- ISO 8-bit code for information interchange -- Structure and rules for implementation (Revision of ISO 4873:1986)	Traitement de l'information -- Code ISO à 8 éléments pour l'échange d'information -- Structure et règles de matérialisation (Revision de l'ISO 4873:1986)	2.1 1.1 89-05
DIS 4902	89-10 2	Information technology -- Data communication -- 37-pole DTE/DCE interface connector and contact number assignments (Revision of ISO 4902:1980)	Communication de données -- Description du connecteur 37 broches à la jonction entre ETTD et ETCD et affectation des broches (Revision de l'ISO 4902:1980)	5.1 4.4 89-06
DIS 4903	89-10 2	Information technology -- Data communication -- 15-pole DTE/DCE interface connector and contact number assignments (Revision of ISO 4903:1980)	Communication de données -- Description du connecteur à 15 broches à la jonction entre ETTD et ETCD et affectation des broches (Revision de l'ISO 4903:1980)	5.1 4.4 89-06
DIS 6522	91-02 2	Programming languages -- General purpose PL/I (Revision of ISO 6522:1985)	Langage de programmation PL/I -- Sous-ensemble pour usage general (Revision de l'ISO 6522:1985)	3.5 3.1 89-06
DIS 7350	89-06 2	Text communication -- Registration of graphic character subrepertoires	Transmission de texte -- Procédure d'enregistrement de sous-répertoires de caractères graphiques	4.4 4.2 89-06
DP 7352	92-06	Information processing systems -- Guidelines for the organization and representation of data elements for data interchange (DTR)	Systèmes de traitement de l'information -- Principes directeurs pour l'organisation et la représentation des éléments de données (DTR)	2.5 2.3 88-08
DIS 7498-4	89-09	Information processing systems -- Open Systems Interconnection -- Basic reference model -- Part 4: Management framework	Systèmes de traitement de l'information -- Interconnexion de systèmes ouverts -- Modèle de référence de base -- Partie 4: Cadre de gestion (DIS distribué en version anglaise seulement)	4.1 3.5 88-08
DP 7776	91-08	Information processing systems -- Data communications -- High-level data link control procedures -- Description of the X.25 LAPB-compatible DTE data link procedures ADDENDUM 1 : PICS proforma	Téléinformatique -- Procédures de commande de liaison de données à haut niveau -- Description des procédures de liaison d'équipement terminal de transmission de données ETTD compatible X.25 LAPB ADDITIF 1 : PICS proforma	2.1 1.1 88-08
DAD 7809	90-11	Information processing systems -- Data communication -- High-level data link control procedures -- Consolidation of classes of procedures ADDENDUM 3: Start/stop transmission	Systèmes de traitement de l'information -- Communication de données -- Procédures de commande de liaison de données à haut niveau -- Consolidation des classes de procedure ADDITIF 3: Transmission commencer/arrêter (DAD distribué en version anglaise seulement)	3.5 3.1 89-03
DAM 7813	90-10	Identification cards -- Financial transaction cards AMENDMENT 1	Cartes d'identification -- Cartes de transactions financieres AMENDEMENT 1 (DAM distribué en version anglaise seulement)	3.5 3.1 89-02
DIS 7816-3	90-04	Identification cards -- Integrated circuit(s) cards with contacts -- Part 3: Electronic signals and transmission protocols	Cartes d'identification -- Cartes à circuit(s) integré(s) à contacts -- Partie 3: Signaux électroniques et protocoles de transmission (DIS distribué en version anglaise seulement)	5.5 5.1 89-06
DP 7826	91-06	Data interchange -- General structure for the interchange of coded representations	Echange de données -- Structure générale pour l'échange de représentations codees	2.8 2.2 88-11
DAD 7942	90-08	Information processing systems -- Computer graphics -- Graphical Kernel System (GKS) functional description ADDENDUM 1	Systèmes de traitement de l'information -- Langage de programmation graphique -- GKS (Graphical Kernel System) description fonctionnelle ADDITIF 1 (DAD distribué en version anglaise seulement)	4.1 3.5 89-06
DAD 8073	89-10	Information processing systems -- Open Systems Interconnection -- Connection oriented transport protocol specification ADDENDUM 2 : Class four operation over connectionless network service	(Actuellement disponible en anglais seulement)	5.1 4.4 89-06

REFERENCE	T A ED	TITLE	TITRE	STAGE AC PR
			(CONTINUED)/(SUITE) JTC 1	
DP 8073	91-05	Information processing systems -- Open Systems Interconnection -- Connection oriented transport protocol specification ADDENDUM 3 : PICS proforma	Systèmes de traitement de l'information -- Interconnexion des systèmes ouverts -- Protocole de transport en mode connexion ADDITIF 3 : PICS proforma	2.2 2.1 89-01
DIS 8208	90-08 2	Information processing systems -- Data communications -- X.25 Packet Level Protocol for Data Terminal Equipment (Revision of ISO 8208:1987)	Systèmes de traitement de l'information -- Communication de donnees -- Protocole X.25 de niveau paquet pour terminal de donnees (Revision de l'ISO 8208:1987) (DIS distribue en version anglaise seulement)	4.1 3.5 89-06
DAD 8208	91-01	Information processing systems -- Data communications -- X.25 packet level protocol for data terminal equipment ADDENDUM 1: Alternative logical channel identifier assignment	Systèmes de traitement de l'information -- Protocole X.25 de niveau paquet pour terminal de traitement de donnees ADDITIF 1: Alternative pour l'attribution d'une reference de voie logique (Distribue en version anglaise seulement)	3.5 3.2 89-05
DAD 8208	90-08	Information processing systems -- Data communications -- X.25 Packet Level Protocol for Data Terminal Equipment ADDENDUM 3: Conformance requirements	Systèmes de traitement de l'information -- Communication de donnees -- Protocole X.25 de niveau paquet pour terminal de donnees ADDITIF 3: Prescriptions de conformite (DIS distribue en version anglaise seulement)	4.1 3.5 89-06
DAD 8326	89-03	Information processing systems -- Open Systems Interconnection -- Basic connection oriented session service definition ADDENDUM 1: Session symmetric synchronization for the session service	Systèmes de traitement de l'information -- Interconnexion de systèmes ouverts -- Définition du service de session de base en mode connexion ADDITIF 1: Synchronisation symétrique de session par le service session (DAD distribue en version anglaise seulement)	4.4 4.1 89-02
DAD 8326	89-10	Information processing systems -- Open Systems Interconnection -- Basic connection oriented session service definition -- ADDENDUM 2: Incorporation of unlimited user data	Systèmes de traitement de l'information -- Interconnexion de systèmes ouverts -- Définition du service de session de base en mode connexion ADDITIF 2: Incorporation de donnees illimitees de l'utilisateur (DAD distribue en version anglaise seulement)	4.4 4.1 89-04
DAD 8326	90-01	Information processing systems -- Open Systems Interconnection -- Basic connection oriented session service definition -- Addendum 3 : Connectionless-mode session service	Systèmes de traitement de l'information -- Interconnexion des systèmes ouverts -- Service de session en mode connexion ADDITIF 3 : Service de session en mode sans connexion (Ce DIS est distribue en version anglaise seulement)	4.1 3.5 88-11
DP 8326	90-04	Information processing systems -- Open Systems Interconnection -- Basic connection oriented session service definition AMENDMENT 1	Systèmes de traitement de l'information -- Interconnexion des systèmes ouverts -- Service de session en mode connexion AMENDEMENT 1	2.1 1.1 88-02
DAD 8327	89-03	Information processing systems -- Open Systems Interconnection -- Basic connection oriented session protocol specification ADDENDUM 1: Session symmetric synchronization for the session protocol	Systèmes de traitement de l'information -- Interconnexion de systèmes ouverts -- Specification du protocole de session de base en mode connexion ADDITIF 1: Synchronisation symétrique de session pour le protocole de session (DAD distribue en version anglaise seulement)	4.4 4.1 89-02
DAD 8327	89-10	Information processing systems -- Open Systems Interconnection -- Basic connection oriented session protocol specification -- ADDENDUM 2: Incorporation of unlimited user data	Systèmes de traitement de l'information -- Interconnexion de systèmes ouverts -- Specification du protocole de session de base en mode connexion ADDITIF 2: Incorporation de donnees illimitees de l'utilisateur (DAD distribue en version anglaise seulement)	4.4 4.1 89-04
DP 8327	91-09	Information processing systems -- Open Systems Interconnection -- Basic connection oriented session protocol specification ADDENDUM 3: Session Protocol PICS Proforma for ISO 8327 and ISO 8327/Addendum 2	Systèmes de traitement de l'information -- Interconnexion des systèmes ouverts -- Protocole de session en mode connexion ADDITIF 3: PICS proforma pour protocole de session selon l'ISO 8327 et l'ISO 8327/Additif 2	2.1 1.1 89-04
DP 8441-1	87-06	High density digital recording (HDDR) -- Part 1 : Unrecorded magnetic tape for HDDR applications	Enregistrement numérique à haute densité (HDDR) -- Partie 1 : Bande magnétique vierge pour les applications HDDR	2.1 1.2 83-10
DP 8441-2	87-06	High density digital recording (HDDR) -- Part 2 : Interchange requirements and test methods for HDDR applications (including the characteristics of recorded magnetic tape)	Enregistrement numérique à haute densité (HDDR) -- Partie 2 : Besoins d'echange et methodes de test pour les applications HDDR (incluant les caracteristiques de la bande magnétique enregistree)	2.1 1.2 83-10
DP 8473	89-10	Information processing systems -- Data communications -- Protocol for providing the connectionless mode network service -- ADDENDUM 2 : Formal description of ISO 8473 (Work suspended)	Additif 2 à l'ISO/DIS 8473 (Titre manque)	2.1 1.2 87-01

CHAPTER 9

NORTH AMERICAN TELECOMMUNICATION STANDARDS ORGANIZATIONS

9.1 INTRODUCTION

This chapter focuses on the telecommunication standards organizations from the United States. The structure of the chapter is to consider first the ISO member organization and then those organizations in government which play a role in telecommunication standards, with the exception of defense. Second, we consider those organizations wholly concerned with telecommunication standards and which are formed to espouse the professional associations', the carriers', and the manufacturers' point of view. Third, we consider organizations which are applications-oriented or user groups and have more of a pressure group role rather than making standards themselves. These groups are more likely to have representatives in other standards organizations.

Chapter 12 considers some North American organizations concerned with functional standardization and specific application areas of standards, for example, in office or manufacturing applications.

Chapter 13 considers further organizations that are international in character, some of which are based in the United States and are concerned with social as well as technical issues. Such organizations tend to be more hybrid in that they consider not only telecommunication *per se*, but also information processing and computerization in its wider forms.

9.2 AMERICAN NATIONAL STANDARDS INSTITUTE

American National Standards Institute (ANSI)
1430 Broadway
New York, NY 10018

Telephone:	212 354 3300
Facsimile:	212 302 1286

The American National Standards Institute (ANSI) was founded in 1918 and has a staff of 110. ANSI is a standards-coordinating and approval organization. The institute does not develop standards. They are developed by qualified technical and professional societies, trade associations, and other groups, which voluntarily submit them to ANSI for approval.

ANSI assigns numbers to standards prepared in accordance with its regulations. These then become national standards. ANSI has established a Joint Telecommunications Standards Coordinating Committee, which coordinates the efforts of the various standards-producing organizations.

Some 8,500 standards approved by ANSI are designated *American National Standards,* which are widely adopted and referenced by government at the federal, state, and local levels.

ANSI accredits certification programs sponsored and operated by competent organizations in the United States, and represents US interests in international nontreaty certification activities. ANSI does not certify products and services.

Some 220 professional and technical societies and trade associations that develop standards in the United States are members of ANSI, as are 1000 companies.

The major functions of ANSI are to

- coordinate the development of voluntary national standards in the United States;
- approve standards as American National Standards;
- manage and coordinate US participation in nongovernmental international standards-developing organizations, such as the International Organization for Standardization (ISO) and the International Electrotechnical Commission (IEC);
- serve as the US source and information center for all American National Standards and those issued by ISO, IEC, and the national standards organizations of 88 other countries; and
- maintain an interface with government at all levels on standards-related matters.

9.3 NATIONAL INSTITUTE OF STANDARDS AND TECHNOLOGY (NIST)

> National Institute of Standards and Technology
> Technology Building 225, Room B151
> Gaithersburg, MD 20899
>
> Telephone: 301 975 2816
> Facsimile: 301 948 1784

In 1968, the National Institute of Standards and Technology (NIST), formerly the National Bureau of Standards, was given responsibility for developing Federal Information Processing Standards (FIPS) for the Federal Automatic Data Processing Community. NIST's National Computer Systems Laboratory (NCSL) in Gaithersburg, Maryland, administers the FIPS program and is the major technical unit within NIST. NCSL primarily focuses on helping the federal government make effective use of computers and information technology. NCSL products, services, and technical support are used by the private sector and all levels of government.

The NCSL develops the FIPS Standard Publications (FIPS PUBS). Over 150 FIPS have been published, including standards, many guidelines, and program information documents. FIPS adopt voluntary industry standards, which are developed with NIST assistance. For example, data encryption was pioneered with the publication of FIPS 46, the data encryption standards. FIPS 46 has been adopted by ANSI as X3.92 and now serves as a basis for further international data encryption work.

NIST (Library Division) is also the secretariat to the National Information Standards Organization (NISO). NISO is the accredited standards developing organization for information, science, libraries, and publishing practices.

9.4 NATIONAL TELECOMMUNICATIONS AND INFORMATION ADMINISTRATION

> National Telecommunications and Information
> Administration
> Department of Commerce
> Washington, DC 20230
>
> Telephone: 202 377 1832
> Facsimile: 202 377 1635

The National Telecommunications and Information Administration (NTIA) was established in 1978 to provide a broad national role in telecommunication and information issues for the Department of Commerce.

NTIA's broad goals include formulating policies to support the development and growth of telecommunication, information, and related industries; furthering the efficient development and use of telecommunication and information services; providing policy and management of the electromagnetic spectrum for federal use, and providing telecommunication facilities grants to public service users.

9.5 US CCITT NATIONAL COMMITTEE

US National Committee for CCITT
Department of State
Bureau of International Communications
and Information Policy
Room 6317 NS
2201 C Street NW
Washington, DC 20520

Telephone: 202 647 5220
Facsimile: 202 647 7407

The participation of the United States in the work of the CCITT is channeled through the national preparatory organization officially known as "The United States Organization for the International Telegraph and Telephone Consultative Committee." This "US CCITT," as it is popularly called, is headed by a director in the Bureau of International Communications and Information Policy, US Department of State.

The US CCITT Charter of 1989, in delineating the purposes of the organization, states that the US CCITT will

- promote the best interests of the United States in CCITT activities;
- provide advice to the Department of State on matters of policy and positions in preparation for CCITT plenary assemblies and meetings of the international CCITT study groups;
- provide advice to the Department of State on the disposition of proposed contributions (documents) to the international CCITT; and
- assist in the resolution of administrative and procedural problems pertaining to US CCITT activities.

Organization and Structure

The US CCITT is composed of a national committee and four study groups. The study group structure does not parallel that of the CCITT itself, but rather each US CCITT study group is concerned with the work of two or more international CCITT study groups. The US CCITT study group structure is flexible and reflects CCITT study-period activities.

The US Committee organization is shown in Table 9.1

Table 9.1. Organization of the US National Committee for CCITT

US Study Group A	*Telecommunications Services and Policy*
Study Group I	Services
Study Group II	Network Operations
Study Group III	Tariff and Accounting Principles
Study Group IX	Telegraph Networks and Telegraph Terminal Equipment
US Study Group B	*Switching, Signaling and ISDN*
Study Group XI	Switching and Signaling
Study Group XVIII	ISDN (Coordinate ISDN related contributions across Study Groups)
US Study Group C	*Telephone Network Operations*
Study Group IV	Maintenance
Study Group V	Protection against Electromagnetic Effects
Study Group VI	Outside Plant
Study Group X	Languages for Telecommunication Applications
Study Group XII	Transmission Performance of Telephone Networks and Terminals
Study Group XV	Transmission Systems and Equipment
US Study Group D	*Data Networks and Telematic Terminal*
Study Group VII	Data Communications Networks
Study Group VIII	Terminals for Telematic Services
Study Group XVII	Data Transmission Over the Telephone Network

The individual study groups organized under US Study Groups A,B,C, and D attempt to reflect the functional organization of the CCITT. The US study groups coordinate their efforts in each CCITT subject area so that they can comprehensively respond to the technical and administrative needs of their members and the CCITT.

Membership in a US CCITT study group is offered to organizations, individuals, operating and manufacturing companies, and user organizations in both the public and private sectors. Broad-based participation in the activities of the US CCITT gives delegates to international CCITT meetings a clearer understanding of the US position on specific telecommunication issues. The formulation of positions is particularly important because the United States is one of the few nations in which the responsibility for and provision of telecommunication services are in the hands of the private sector with only governmental oversight and limited direct involvement.

According to the US CCITT Charter, the dates of each US CCITT study group meeting and proposed agenda must be announced to the public in both the Federal Register and a Department of State press release at least 15 working days prior to the convening of the meeting. These meetings are open to the public.

By participating in US CCITT activities, non-CCITT telecommunication organizations and individuals have a channel for influence on contributions to the CCITT. Also, as the formal CCITT meeting reports may be received from Geneva as late as three to six months following a meeting, attendance at the US CCITT meetings permits participants to keep abreast of the CCITT deliberations on emerging technology.

Each CCITT study group has three or four plenary meetings that involve the entire study group within the four-year study period between Plenary Assemblies. Generally, a CCITT study group is subdivided into working parties, each led by a chair and responsible for a subset of related items among the questions to be studied.

Contributions to the CCITT from the United States are of two types, classified by the source of the document: *individual-member contributions,* and *US contributions.* Generally, the US contribution is either

- an individual-member contribution approved by the appropriate US CCITT study group as representing an agreed US contribution; or
- a contribution from some active CCITT member of the US CCITT study group, such as an individual firm, government agency, or national standards group or committee (for example, a Technical Committee of the American National Standards Institute (ANSI); an approved Accredited Standards Committee, such as X3, "Information Processing Systems," or T.1, "Telecommunications").

9.6 FEDERAL COMMUNICATIONS COMMISSION

Federal Communications Commission
1119 M Street NW
Washington, DC 20554

Telephone: 202 632 6000
Facsimile: 202 653 5402

Origins and Structure of the Federal Communications Commission

The Communications Act of 1934, as amended (47 USC 151) established the Federal Communications Commission (FCC) to regulate interstate and foreign commerce in communication by wire and radio "in the public interest."

The FCC is composed of five members, called Commissioners, who are appointed by the President with the approval of the Senate. One of the members is designated as Chairman by the President.

In administering the programs necessary to its regulatory responsibility, the FCC is assisted by a General Counsel, who, in addition to typical duties, exercises exclusive control of court appeals involving broadcast matters; a Managing Director; a Director of Congressional and Public Affairs; a Chief Engineer; a Chief of Plans and Policy; and the chiefs of four bureaus to whom certain licensing and grant authority is delegated.

To assist the FCC in the adjudicatory process, there is a Review Board to review initial decisions and to write decisions, and an Adjudication Division in the office of the General Counsel to assist the FCC and individual Commissioners in the disposition of cases in adjudication (as defined in the Administrative Procedure Act (5 USC note prec.551) that have been designated for hearings. There also is a corps of administrative law judges (qualified and appointed pursuant to the requirements of the Administrative Procedure Act), who conduct evidentiary hearings and write initial decisions.

Qualification Certification

Applicants wishing to establish a radio station, whether it be broadcast, private, common carrier, or experimental, must submit information showing compliance with the FCC Rules. If the requirements (standards) are met, a license is issued. As of 1980, the FCC had issued approximately 1.5 million radio licenses. The FCC also has four equipment authorization programs, which issue equipment approvals each year.

The FCC participates in organizations and committees that deal with telecommunication standards, such as the Institute of Electrical and Electronics Engineers, the American National Standards Institute, and the Electronic Industries Association.

Standardization Activities

The FCC participates in international and national standards organizations developing communication systems and equipment, and provides technical analysis and testing of them.

Mass Media

The Mass Media Bureau administers the regulatory program for

- direct broadcast satellite (DBS), standards, frequency modulation (FM), television, low power television (LPTV), translators, instructional television, fixed services (FS), related broadcast auxiliary services, cable television (CATV), microwave radio relay, and registration of new cable television systems.

The bureau issues construction permits, operating licenses, and renewals or transfers of such licenses, and processes registrations, notifications, and petitions for CATV systems.

The bureau also oversees compliance by broadcasters and CATV system operators with statutes and FCC policies, and is responsible for maintaining relations with state and local authorities, who share responsibility for the regulation of cable systems.

Common-Carrier Communication

The Common Carrier Bureau administers the regulatory program in interstate and international common-carrier communication by telephone, telegraph, radio, and satellite. Common carriers include companies, organizations, or individuals providing communication services to the public for hire, and must serve all who wish to use the services at established rates. In providing interstate and foreign communication service, common carriers may employ landline wire, electrical or optical cable facilities, point-to-point microwave radio (signals relayed by stations spaced at given intervals), land mobile radio (two-way telephone or one-way signaling communication between base and mobile units), cellular radiotelephone, or satellite systems. Communication services between the United States and overseas points by common carriers are provided by transoceanic cable, high frequency (HF) radio, and satellite communication.

Private Radio Communication

The Private Radio Bureau regulates the use of the radio spectrum to fulfill the communication needs of businesses, state and local governments, aircraft, ships, and individuals. Over 2.5 million licensees use radio to promote safety of life and property, to increase productivity, and to advance the science of telecommunication. The Private Radio Bureau regulates and licenses two broad groups of radio services:

- the private land mobile and microwave radio services, and
- the special radio services.

Private land mobile and microwave radio services are used by public safety (police, fire, local government), industrial (businesses including public utilities), and land transportation entities (buses, railroads, taxicabs).

Special radio services include aviation, the marine, the Alaska fixed service, and personal radio services (amateur and general mobile). While two personal radio services, citizens (CB) and radio control (R/C), are regulated, licensure is not required. The bureau also implements the compulsory provisions of laws and treaties covering the use of radio for the safety of life and property at sea and in the air.

9.7 INSTITUTE OF ELECTRICAL AND ELECTRONICS ENGINEERS (IEEE)

Institute of Electrical and Electronics Enigineers
345 East 47th Street
New York, NY 10017-2394

Telephone: 212 705 7900
Facsimile: 212 752 4929

Origins, Membership, and Purpose

The Institute of Electrical and Electronics Engineers was formed in 1963 by a merger of the American Institute of Electrical Engineers (founded in 1884) and the Institute of Radio Engineers. The IEEE is a transnational, individual-member society. During its more than 100-year history, IEEE has grown to become the world's largest technical professional organization with 300,000 members in over 130 countries. IEEE's objectives are scientific, eductional, professional, and societal. Its technical objectives are advancing the theory and practice of electrical, electronic, and computer engineering.

The IEEE's 36 societies (subdivisions by technical specialty) sponsor some 250 major technical conferences each year. Each member of the IEEE has the opportunity to join any number of societies, which are grouped administratively into 10 divisions as indicated in Table 9.2. The Computer Society is the largest.

In countries other than the United States, the IEEE works in cooperation with the appropriate national society. In the United States, programs are implemented by the United States Activities Board, headquartered in Washington, DC.

Organization

The IEEE's governing body, the Board of Directors, comprises 32 volunteer directors and officers, plus the Executive Director. Twenty-four of its members are elected directly by the voting membership. This group, meeting separately as the Assembly, then elects the remaining nine members, including seven officers, who complete the board. Twenty members of the Assembly include one delegate director from each of the ten IEEE world geographic regions, and one from each of the ten technical divisions. The remaining four are the Institute President, Executive Vice President, President-Elect, and Past President.

A 12-member Executive Committee of the Board of Directors has the task of implementing the board's policies and operating the institute with the assistance of the staff.

Activities are overseen by various volunteer boards and committees, including those on awards, educational activities, publications, regional activities, standards, technical activities, and United States activities.

The IEEE is headquartered in New York. Other major locations are the service center in Piscataway, New Jersey, from which its many publications can be ordered, and the United States activities office in Washington, DC. The IEEE Computer Society, with offices in Washington, DC, and Los Altos, California, also has offices in Brussels, Belgium, and Tokyo.

Standards Activities

One of the IEEE's missions is to gather, organize, and dissseminate techni-

Table 9.2. IEEE Divisions and Societies (from *IEEE Spectrum,* ©1990 IEEE)

Division I—Circuits and Devices
        Circuits and Systems
        Components, Hybrids, and Manufacturing Technology
        Electron Devices
        Lasers and Electro-Optics
        Solid-State Circuits Council *
Division II—Industrial Applications
        Dielectrics and Electrical Insultation
        Industry Applications
        Instrumentation Measurement
        Power Electronics Society **
Division III—Communications Technology
        Broadcast Technology
        Communications
        Consumer Electronics
        Vehicular Technology
Division IV—Electromagnetics and Radiation
        Antennas and Propagation
        Electromagnetic Compatibility
        Magnetics
        Microwave Theory and Techniques
        Nuclear and Plasma Sciences
Divisions V and VIII—Computer
        Computer
Division VI—Engineering and Human Environment
        Education
        Engineering Management
        Professional Communication
        Reliability
        Social Implications of Technology
Division VII—Power Engineering
        Power Engineering
Division IX—Signals and Applications
        Acoustics, Speech, and Signal Processing
        Aerospace and Electronic Systems
        Geoscience and Remote Sensing
        Oceanic Engineering
        Ultrasonics, Ferroelectrics, and Frequency Control
Division X—Systems and Control
        Control Systems
        Engineering in Medicine and Biology
        Industrial Electronics
        Information Theory
        Robotics and Information Council *
        Systems, Man, and Cybernetics

*Council members consist of Societies, not individuals.
**As of January 1, 1988, Power Electronics changed from a Council to a Society.

cal information; standards help to complete that task. Some IEEE publications, such as *Transactions* and *Proceedings,* present individual opinions and differences of opinion on particular subjects. Standards complement that process by presenting a consensus of thought on a particular issue. Some 600 IEEE standards have been published and recommended for use, and every year approximately 60 new or revised standards are developed. Once a need for a standard is clearly defined, a detailed procedure is followed, which involves review and approval by several different committees. A voluntary process, standards development uses a consensus approach, including participation of all interested parties. Proposed standards are scrutinized by a review committee to ensure that due process is observed, a proper balance of interested parties exist, and procedural correctness of coordination and balloting has been maintained.

Standards

The IEEE produces an "IEEE Standards Submittal Kit," comprising six documents:

1. Standards Project Authorization (PAR) Submitters Working Guide;
2. Working Guide for Submittal of Proposed Standards and Form for Submittal of Proposed Standards;
3. Approved Project Authorization Request and Ballot Summary;
4. IEEE Standards Manual;
5. A Guide to IEEE Standards Development;
6. IEEE's Standards Style Manual.

To follow are Table 9.3 on IEEE organization, Table 9.4 on a possible standards development structure, and Table 9.5 on the standards development process.

The IEEE Standards Manual is the rule book followed in the development of IEEE standards. To some extent, the IEEE's manual mirrors the ANSI document, the American National Standards Procedures for the Development and Co-ordination of American National Standards. The IEEE Standards Board is currently consolidating and reorganizing such manuals.

The IEEE Standards Board is responsible throughout the institute for encouraging and coordinating the development and revision of IEEE stand-

Table 9.3. IEEE Organization (from *IEEE Spectrum,* ©1990 IEEE)

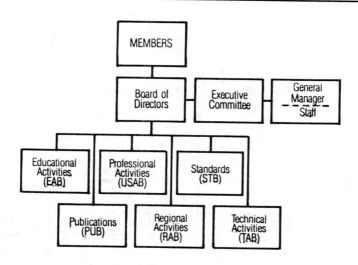

Table 9.4. Example of a Possible Standards Development Structure (from *A Guide to IEEE Standards Development,* ©1990 IEEE)

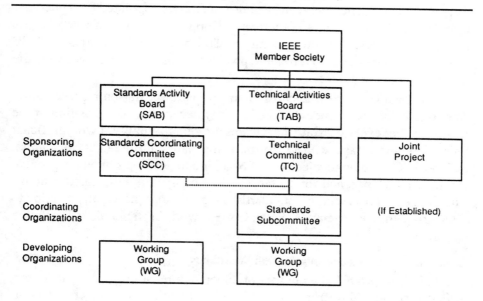

Table 9.5. IEEE Standards Development Process (from *A Guide to IEEE Standards Development,* ©1990 IEEE)

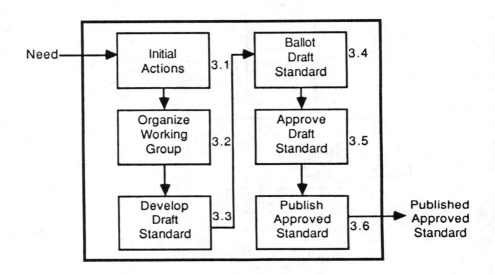

ards, and for other standards-related activities in the organization's field of interest. The board has between 18 and 26 members appointed for a one-year term by the Nomination Appointments Committee and confirmed by the Board of Directors. The voting members of the board are from industry, IEEE technical societies, and government, providing a balanced perspective in the development of standards.

The bylaws of the Standards Board provide for liaison representation from organizations outside the IEEE, partly because the composition of the board changes every year and the liaisons provide continuity. Two significant liaison organizations are NIST (formerly the National Bureau of Standards), and the FCC. Principal objectives of NIST are to develop standards, provide technical assistance, and conduct research on computers and related systems. The FCC develops technical standards and guidelines principally relating to radio-frequency interference (RFI). Current work in standards is illustrated by Table 9.6.

Computing and Telecommunication Standards

The IEEE has produced a number of telecommunication-related standards, primarily through the efforts of the IEEE Communications Society's techni-

Table 9.6. Standards Action of IEEE Standards Board, November 9, 1989 (from *IEEE Standards Bearer,* Vol. 3, No. 5, December 1989, ©1989 IEEE)

■ **Newly Approved Project Authorization Requests (PARs)**

■ **Approved New Standards**

■ **Approved Standards Revisions, Reaffirmations, and Withdrawals**

■ **APPROVED PARs FOR NEW STANDARDS**

P1014.1 VMEbus, Futurebus+Extended Architecture (VFE)

P1226 ATLAS/Ada Based Environment for Test (ABET)

P1227 Guide for the Measurement of DC Electric Field Strength and Ion Related Quantities

P1228 Standard for Software Safety Plans

P1229 Definitions for Electrical Heat Tracing

P1230 Guide for Measurement of Radio Noise from the Testing of High Voltage Hardware and Insulators

P1296.1 Interfacing Between IEEE Std 1296 (Multibus II) and IEEE Std 896 (Futurebus+)

PC37.40c Definitions for Full Range Current Limiting Fuse

PC37.41h Design Tests for Full Range Current Limiting Fuse

PC37.48d Application, Operation, and Maintenance Guidelines for Full Range Current Limiting Fuse

PC57.19.03 Standard Requirements, Terminology, and Test Code for Bushings for DC Applications

■ **APPROVED PARs FOR STANDARDS REVISIONS**

P18 Standard for Shunt Power Capacitors

P200 Reference Designations for Parts and Equipment

P622A Design and Installation of Electric Heat Tracing Control and Alarm Systems for Power Generating Stations

PC57.19.01 Standard Performance Characteristics and Dimensions for Outdoor Apparatus Bushings

■ **REVISED PARs**

P896.1 Futurebus–Logical Protocol Specification

P896.2 Futurebus–Physical Specifications and Profiles

P896.3 Futurebus–Systems Configuration

P1095 Guide for Installation of Vertical Generators and Generator/Motors for Hydroelectric Applications

P1120 Guide to the Factors to Be Considered in Planning, Design, and Installation of Submarine Power and Communication Cable

PC62.1 Standard for Gapped Silicon-Carbide Surge Arresters for AC Power Circuits

770X3.160 Standard for the Programming Language Extended Pascal

■ **WITHDRAWN PARs**

P896.4 Futurebus+System Configuration Guide

P942 A Semiconductor Device Test Programming Language

P1192 Standard for a Microcomputer Database Language

■ **APPROVAL OF NEW STANDARDS**

857 Guide for Test Procedures for HVDC Thyristor Valves

*1095 Guide for Installation of Vertical Generators and Generator/Motors for Hydroelectric Applications

■ **APPROVAL OF REVISED STANDARDS**

*48 Test Procedures and Requirements for High-Voltage Alternating Current Cable Termination

*770X3.160 Standard for the Programming Language Extended Pascal

C62.1 Standard for Gapped Silicon-Carbide Surge Arresters for AC Power Circuits

■ **APPROVAL OF REAFFIRMED STANDARDS**

521 Letter Designations for Radar Frequency Bands

765 Preferred Power Supply for Nuclear Power Generating Stations

*Final approval date subject to all Standards Board conditions being met.

cal committees. The Communications Society has 17 technical committees. About 10 of these are actively involved in developing standards. Some of the more active committees are the Data Communications Systems Committee, Transmission Systems Committee, Telecommunication Switching Committee, Computer Communications Committee, Standards Coordinating and Liaison Committee, and Subcommittee on Telecommunications Measurement Standards. The Standards Coordinating and Liaison Committee brings together all standards activities of the Communications Society.

IEEE processes for developing standards are

1. All IEEE standards activities are under the direction of the IEEE Standards Board. Standards are developed by the various technical committees of the IEEE societies.
2. An IEEE Standards Manual and an IEEE Standards Style Manual list the procedures for new standards projects, the formats to be used, and the procedures for submitting completed standards to the Standards Board for approval.
3. The Standards Board has established a number of Standards Coordinating Committees (SCCs) for various IEEE activities. SCC-25 is particularly active in coordinating all telecommunication standards activities of the various societies within IEEE.
4. New Standard Project Authorization Requests must be submitted to the Standards Board for approval before any long-term work on the standard begins.
5. All standards completed by a committee must be submitted to the Standards Board with the results of a vote by all committee members and a summary of the committee's make-up to ensure that the work was done by a balanced representation of the industry involved.
6. Published standards are submitted to ANSI by the IEEE Standards Office for consideration as national standards.
7. The IEEE Communications Society (COMSOC) also has a Standards Liaison and Coordinating Committee, which brings together all COMSOC standards activities to ensure compliance with the IEEE Standards Manual and to avoid duplication of effort by the various COMSOC technical committees.

Publications

More than 20% of the world's literature in electrotechnology is published by the IEEE, including the monthly magazine *IEEE Spectrum,* which goes to all members. In total, the IEEE publishes and distributes some 70 specialized journals and magazines, 130 conference records, and a variety of IEEE Press books.

Two publications, *IEEE Spectrum* and *Proceedings of the IEEE,* treat technical subject mattter in such a way that it is comprehensible and informative, not only to the specialist, but also to the general IEEE member. *Proceedings of the IEEE* emphasizes papers providing in-depth reviews and tutorial coverage in fields of interest. The IEEE produces a publications catalogue, which includes a list of all periodicals published by the IEEE.

All of the IEEE's specialized periodicals are sponsored by one or more of the individual societies. The *Transactions* (journals) appear as quarterly or monthly publications that concentrate on theoretical or experimental papers. The magazines are applications-oriented. Relevant journals and magazines follow.

IEEE Transactions on Computers
IEEE Computer Magazine
IEEE Network
IEEE Transactions on Acoustics, Speech, and Signal Processing
IEEE ASSP Magazine (Practical applications of acoustics, speech, and signal processing)
IEEE Journal on Selected Areas in Communications
IEEE Transactions on Communications
IEEE Communications Magazine

The IEEE also publishes a book of standards. Some relevant telecommunication standards are the local area network (LAN) series 802.2 to 802.5.

For information,

IEEE, Washington Office
1111 19th Street NW
Washington, DC 20036-3690

| Telephone: | 212 785 0017 |
| Information: | 202 785 2180 |

For publications,

IEEE Service Center
445 Hoes Lane
Piscataway, NJ 08855-1331

Telephone:	201 981 0060
Sales Information:	201 981 1393

9.8 EXCHANGE CARRIERS STANDARDS ASSOCIATION (ECSA)

Exchange Carriers Standards Association
Suite 200
5340 Grosvenor Lane
Bethesda, MD 20814-2122

Telephone:	301 564 4505
Facsimile:	301 564 4501

The ECSA is a trade association of wireline exchange carriers representing, more than 95% of all telephone subscribers in the United States.

The ECSA funds the staff which serves as the T1 Secretariat. ECSA's Standards Advisory Committee (SAC) is responsible to ANSI for its sponsorship of T1. In addition, the SAC oversees the T1 secretariat, ensures T1's conformance with ANSI procedures, and provides T1 with the independence to operate as an open committee serving the industry at large.

The Exchange Telephone Group Committee (ETGC) of ECSA provides a forum for and representation of exchange carrier interests in standards and related matters. The ETGC has assumed the role of the former telephone group as the exchange carrier representative within ANSI. The ETGC draws upon ECSA member organizations and their affiliates to develop positions and to provide representation on numerous ANSI boards and accredited committees. A liaison committee represents ECSA in developing a coordinated presence for ECSA within the standards and exchange carrier communities.

9.9 STANDARDS COMMITTEE T1

Exchange Carriers Standards Association
Standards Committee T1
Suite 200
5340 Grosvenor Lane
Bethesda, MD 20814-2122

Telephone: 301 564 4505
Facsimile: 301 564 4501

Standards Committee T1 provides a public forum for developing interconnection standards for the nationwide telecommunication network.

To ensure that standards projects represent an industry consensus, T1 has been accredited by ANSI. The name T1, not to be confused with the well-known T-1 transmission system, is attributed to ANSI's coding of standards committees: T for telecommunication and 1 for the first such ANSI entity.

In August 1983, ECSA recommended to the FCC that a public standards committee be established, which would be open in membership and would use the same ANSI procedures as countless other existing standards committees. Although it develops model procedures for standards-setting, ANSI does not sponsor standards activities. While ANSI coordinates standards activities, it views the responsibility of sponsorship as belonging to elements of the industry that implements the standards. Consequently, the FCC was advised by ECSA that it was prepared to sponsor an ANSI-affiliated committee and to provide the secretariat and administrative functions for the committee's work.

Following ANSI practices, committee members are classified by their principal interest to determine whether a cross section of relevant interests are involved in the process. Four broad interest categories have been defined:

- exchange carriers;
- interexchange carriers and resellers;
- manufacturers and vendors; and
- general interests, which include government agencies, consultants, user groups, and liaison with other committees.

The committee structure provides for a number of specialized technical subcommittees. Membership procedures allow for organizations to join specific subcommittees without requiring membership in the full committee. In addition, all meetings are open to the public. Table 9.7 shows the way in which proposed standards are made by the committee.

The organizational interfaces are illustrated in the table. Projects flow into T1, where they are addressed within a structure of technical subcommittees. Projects intended to yield candidate American National Standards will flow to ANSI and its Board of Standards Review. In addition, T1's work program also includes liaison projects related to continuing work in international forums, such as the CCITT. T1 expects to formulate industry positions on CCITT-related topics and to submit these to the US National Committee for CCITT. A third aspect is the presentation of industry reports.

Table 9.7. Standards Committee T1 Project Flow (from *IEEE Communications Society Magazine,* Vol. 23, No. 1, January 1985, ©1985 IEEE)

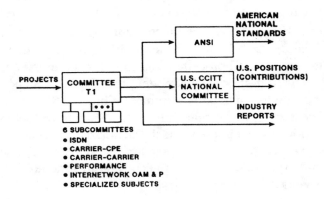

T1 Committee Structure

The T1 structure features a main committee, an advisory group, and six technical subcommittees. Each subcommittee establishes working groups, as needed, to address specific projects. All elements of T1 are subordinate to the main committee. The committee establishes the organizational structure, elects officers, approves standards projects, and approves candidate standards for submission to ANSI.

The advisory group (known as T1AG) is an executive committee, which manages the main committee's activities between T1's general meetings. The six Technical Subcommittees and their work programs are outlined below:

Network Interfaces (TIE1)

- Analog Access (T1E1.1)
- Wideband Access (T1E1.2)
- Connectors and Wiring Arrangements (T1E1.3)
- DSL Access (T1E1.4)
- Editorial (T1E1.5)

Internetwork Operations, Administration, Maintenance (OAM), and Provisioning (T1M1)

- Internetwork Planning and Engineering (T1M1.1)
- Internetwork Operations (T1M1.2)

- Testing and Operations Support Systems and Equipment (T1M1.3)
- Administrative Systems (T1M1.4)
- OAM and P Architecture, Interfaces and Protocols (T1M1.5)

Performance (T1Q1)

- Voice and Voice-Band Data (T1Q1.1)
- Digital Packet and ISDN (T1Q1.3)
- Digital Circuit (T1Q1.4)
- Wideband Program (T1Q1.5)

Services, Architecture, and Signaling (T1S1)

- ISDN Architecture and Services (T1S1.1)
- ISDN Switching and Signaling Protocols (T1S1.2)
- Common Channel Signaling (T1S1.3)
- Numbering and Individual Channel Signaling (T1S1.4)
- Broadband ISDN (T1S1.5)

Digital Hierarchy and Synchronization (T1X1)

- Synchronization Interfaces (T1X1.3)
- Metallic Hierarchical Interfaces (T1X1.4)
- Optical Hierarchical Interfaces (T1X1.5)
- Tributary Analysis (T1X1.6)

Specialized Subjects (T1Y1)

- Specialized Video and Audio Services (T1Y1.1)
- Specialized Voice and Data Processing (T1Y1.2)
- Environmental Standards for Exchange and Interexchange Carrier Network (T1Y1.4).

9.10 ELECTRONIC INDUSTRIES ASSOCIATION (EIA)

Electronic Industries Association
Standards Sales
2001 I Street NW
Washington, DC 20006

Telephone: 202 457 4966
Facsimile: 202 457 4985

The Electronic Industries Association (EIA) was founded as the Radio Manufacturers Association in 1924. EIA is a prominent standards-development group in the field of data communication. Membership includes manufacturers of small components as well as large corporations that design and produce complex systems for defense, space, and industry. As a trade organization, the EIA is involved in standards activities.

The comprehensive standards programs, include over 600 standards produced by more than 5,000 industry and government participants in some 300 committees. The EIA is an ANSI Accredited Standards Organization, and the EIA Committees contribute to international standards of the ISO, CCITT, and IEC. Their best known standard is the EIA-232-C computer-printer interface (formerly RS-232-C), which is widely used throughout the world.

In 1988, the telecommunication sector of EIA merged with the US Telecommunication Suppliers Association (USTSA) to form a new group, the Telecommunication Industry Association (TIA). The new organization operates in association with EIA and continues to act as the telecommunication sector of EIA. Telecommunication standards work sponsored by TIA is done through the EIA Standards Department, and future telecommunication standards issued by this group will be identified as "EIA/TIA."

Active telecommunication standards committees operating under the TIA-EIA structure are

TR-80	Mobile Radio
TR-14	Microwave Radio
TR-29	Facsimile Systems and Equipment
TR-30	Data Transmission Systems and Equipment
TR-32	Personal Radio
TR-34	Satellite Equipment and Systems
TR-41	Telephone Terminals
TR-45	Cellular Radio
FO-2	Optical Communications
FO-6	Fiber Optics

9.11 INTERNATIONAL COMMUNICATIONS ASSOCIATION (ICA)

International Communications Association
Suite 710 LB-89

12750 Merit Drive
Dallas, TX 75251

Telephone: 214 233 3889
Facsimile: 214 233 2813

The International Communications Association (ICA) was formed in 1948 in response to the increasing importance of telecommunication as an integral part of the business world. ICA is a full member of the International Telecommunication Users' Group (INTUG) and is the largest telecommunication users' group in the United States. ICA has more than 1,500 members, representing 600 organizations from industry, commerce, government, and education. For membership, organizations must have telecommunication expenditures of $1 million or more per year. Collectively, ICA members spend approximately $15 billion per year on communication equipment and services.

The main purpose of the association is to be the most comprehensive and timely source for telecommunication information and education. This range of information and education covers voice, data, and image communication, as well as the interrelationships among the disciplines. ICA acts as a forum for members to exchange ideas and experiences in the communication field, sets professional standards, provides vendor and regulatory liaison, presents awards and scholarship grants, and holds interim seminars as well as an annual conference and exposition.

9.12 THE COMPUTER AND BUSINESS EQUIPMENT MANUFACTURERS' ASSOCIATION (CBEMA)

The Computer and Business Equipment
Manufacturers' Association
311 First Street NW #500
Washington, DC 20001

Telephone: 202 737 8888
Facsimile: 202 638 4922

The Computer and Business Equipment Manufacturers' Association (CBEMA) represents the interests of leading companies providing computer and business equipment and telecommunication hardware, software, and services. CBEMA focuses on

• public policy issues beneficial to the information technology industry;

- standards;
- industry councils; and
- international liaison.

CBEMA's mission has 19 goals, which include

- developing consensus among members on the actions and policies necessary to support the long-range vitality of the computer, business equipment, and telecommunication industries;
- maintaining the CBEMA's leadership role in standards and striving to optimize standards in the United States, leading to harmonization with international standards;
- improving the voluntary standards organizations and processes (domestically and internationally) in terms of usefulness, time, cost, and quality.

Organization

CBEMA's direction is controlled and organized by

- a board of directors;
- an executive committee;
- a plans and program committee;
- a program and budget committee;
- a long-range planning committee; and
- subcommittees of the board of directors.

The committee structure is shown in Table 9.8. The internal organization is shown in Table 9.9.

Relevant committees include the International Committee, which was involved in 1989 after the United States Trade Representative determined that Japan had violated the Market-Oriented Sector Specific (MOSS) Talks Telecommunication Agreement, which sets out market access opportunities in the telecommunication area. Eventually, significant progress was made in efforts to open the Japanese telecommunication market in the areas of cellular telephone and third-party radio.

CBEMA also participated in the International Information Industry Congress (IIIC) at its recent meeting in Canberra, Australia, where CBEMA presented views on GATT initiatives, intellectual property rights, trade in services, and foreign investment.

Table 9.8. CBEMA's Committee Structure

```
                        ┌─────────────┐
                        │   Board     │
                        │   of        │
                        │   Directors │
                        └─────────────┘

┌────────────┐  ┌────────────┐  ┌────────────┐  ┌──────────────┐
│ Nominating │  │ Executive  │  │ Long Range │  │ Performance &│
│ Committee  │  │ Committee  │  │ Planning   │  │ Compensation │
│            │  │            │  │ Committee  │  │ Committee    │
└────────────┘  └────────────┘  └────────────┘  └──────────────┘

                        ┌─────────────┐
                        │ Plans &     │
                        │ Programs    │
                        │ Committee   │
                        └─────────────┘

┌──────────┐ ┌────────────┐ ┌────────────┐ ┌────────────┐ ┌──────────┐
│ Program  │ │ Industry   │ │ Educational│ │ Public     │ │ Industry │
│ & Budget │ │ Committees │ │ Councils   │ │ Policy     │ │ Support  │
│ Committee│ │            │ │            │ │ Committee  │ │          │
└──────────┘ └────────────┘ └────────────┘ └────────────┘ └──────────┘
```

Industry Committees	Educational Councils	Public Policy Committee	Industry Support
Standards Program Management Committee	Chief Tax Executives	Committee for Action on Public Policy	Industry Marketing Statistics Committee
Environment & Safety Management Committee	Export Controls	Government Procurement	Communications Committee
Ergonomic Committee	Human Resources	International Issues	
CBEMA/UL Liaison Committee	Logistics	Proprietary Rights	
Top Vendors Group	Service Management	Tax	

Table 9.9. CBEMA's Internal Organization

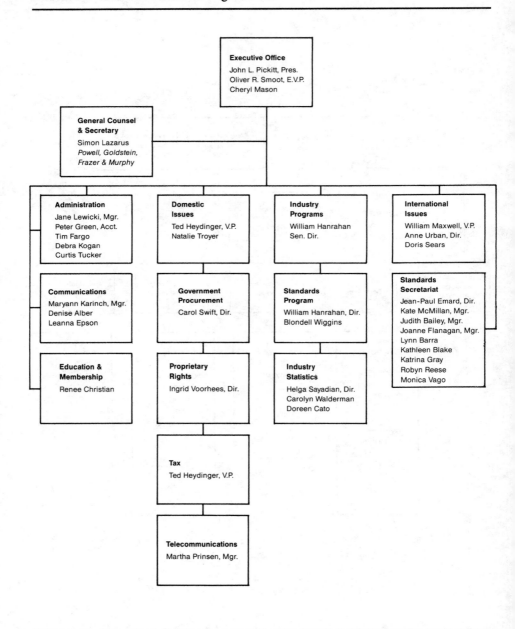

Executive Office
John L. Pickitt, Pres.
Oliver R. Smoot, E.V.P.
Cheryl Mason

**General Counsel
& Secretary**
Simon Lazarus
*Powell, Goldstein,
Frazer & Murphy*

Administration
Jane Lewicki, Mgr.
Peter Green, Acct.
Tim Fargo
Debra Kogan
Curtis Tucker

Communications
Maryann Karinch, Mgr.
Denise Alber
Leanna Epson

**Education &
Membership**
Renee Christian

**Domestic
Issues**
Ted Heydinger, V.P.
Natalie Troyer

**Government
Procurement**
Carol Swift, Dir.

**Proprietary
Rights**
Ingrid Voorhees, Dir.

Tax
Ted Heydinger, V.P.

Telecommunications
Martha Prinsen, Mgr.

**Industry
Programs**
William Hanrahan
Sen. Dir.

**Standards
Program**
William Hanrahan, Dir.
Blondell Wiggins

**Industry
Statistics**
Helga Sayadian, Dir.
Carolyn Walderman
Doreen Cato

**International
Issues**
William Maxwell, V.P.
Anne Urban, Dir.
Doris Sears

**Standards
Secretariat**
Jean-Paul Emard, Dir.
Kate McMillan, Mgr.
Judith Bailey, Mgr.
Joanne Flanagan, Mgr.
Lynn Barra
Kathleen Blake
Katrina Gray
Robyn Reese
Monica Vago

The Standards Program Management Committee (SPMC) implemented a study of the effectiveness of the US Information Technology Voluntary Standards Program from the business perspective. In this committee, CBEMA worked with its counterpart from Canada, the Information Technology Association of Canada (ITAC), focusing on harmonizing relevant North American standards.

CBEMA also participated in the expansion of the IEC System for Conformity Testing to Standards for Safety of Electrical Equipment (IECEE). The IECEE system will determine the requirements for certification of the electrical safety of US products in the European Community, European Free Trade Agreement Area, and in other countries significant to US trade. Through membership in the IECEE, US information technology manufacturers should be able to have all their products tested one time only in the United States, and then be able to market them in all other IECEE member countries.

Standards

The standards program operates under the guidance of the Standards Program Management Committee (SPMC) and interacts with all important standards developing organizations worldwide. CBEMA's standards work is depicted in Table 9.10. CBEMA does not develop or promulgate standards, but the members actively participate in developing them.

CBEMA serves as the secretariat of standards committee X3 (Information Processing Systems). X3 is accredited by ANSI to develop voluntary American National Standards in the areas of media, programming, languages, documentation, systems, and intercommunication among computing devices and systems. CBEMA has served as the secretariat since X3 was formed by ANSI in 1960.

CBEMA has also been appointed by ANSI as the technical administrator for the US Technical Advisory Group (TAG), which develops US contributions to and positions on international standards, covering information technology on ISO and IEC.

As secretariat for X3, CBEMA has been responsible for producing a manual for all projects currently under review, the *X3 Projects Manual*.

9.13 STANDARDS COUNCIL OF CANADA (SCC)

Standards Council of Canada
Suite 1200, 350 Sparks Street

Table 9.10. CBEMA's Industry Programs

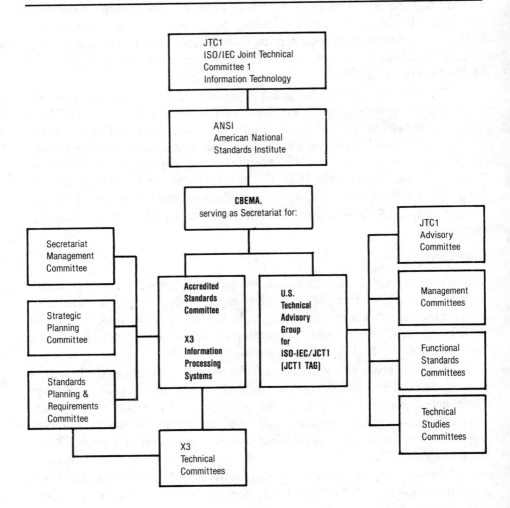

Ottawa, Ontario K1P 6N7
Canada

Telephone:	613 238 3222
Facsimile:	613 995 4564

The Standards Council of Canada (SCC) was founded in 1970. The SCC is a crown corporation and the national coordinating agency for standardization, bringing together, into a National Standards System, established organizations involved in the preparation of voluntary standards for application in both the private and public sectors, testing, and certification. The National Standards System is committed to a continuing program of processing existing and new Canadian standards for approval as National Standards by the Standards Council of Canada.

The SCC has a 69-member staff directly employed by the member body. The objectives of the SCC, as set forth by the statute, are to

- foster and promote voluntary standardization in fields relating to the construction, manufacture, production, quality, performance and safety of buildings, structures, manufactured articles and products and other goods, including components thereof, not expressly provided for by law, as a means of advancing the national economy, benefiting the health, safety and welfare of the public, assisting and protecting consumers, facilitating domestic and international trade and furthering international cooperation in the fields of standards.

The primary responsibilities of the SCC include sale of publications, education, and promotion.

CHAPTER 10

EUROPEAN TELECOMMUNICATION STANDARDS ORGANIZATIONS

10.1 INTRODUCTION

The European countries considered in this chapter are Austria, Belgium, Denmark, the Federal Republic of Germany (FRG), Finland, France, Greece, Iceland, Ireland, Italy, Luxembourg, the Netherlands, Norway, Portugal, Spain, Sweden, Switzerland, and the United Kingdom.

This chapter focuses on the European participants in telecommunication standardization. We discuss

- the European Community's governing institutions;
- telecommunication standards organizations;
- current procedures for standards-making in the European Community;
- implementation of the *Green Paper;*
- other telecommunication organizations affecting standards-making;
- programs involving telecommunications; and
- major European members of ISO.

10.2 GOVERNING INSTITUTIONS IN THE EUROPEAN COMMUNITY

Legal Status, Members, Aims

The European Community (EC) is a system comprising three distinct entities:

- European Coal and Steel Community (ECSC);
- European Economic Community (EEC); and
- European Atomic Energy Community (Euratom).

The ECSC was established by the Treaty of Paris in 1952, and the EEC and Euratom were established in 1958 by the Treaty of Rome. On 1 July 1987, the Single European Act gave formal legal status to European political cooperation.

Members of the EC are Belgium, Denmark, France, FRG, Greece, Ireland, Italy, Luxemburg, the Netherlands, Spain, Portugal, and the United Kingdom.

The EC was formed to promote a harmonious development of economic activities by the promotion of closer relations among the member states, establishing a common market, and progressive rapprochement of the economic policies of member states.

The common institutions to the three communities are

- the European Parliament (or Assembly);
- the Council of Ministers (or European Council);
- the Commission of the European Communities (CEC);
- the Court of Justice; and
- the Court of Auditors.

The European Parliament

The European Parliament is an assembly of 518 EEC representatives of each member country. The number of representatives for each country are

France	81	Belgium	24
Germany	81	Greece	24
Italy	81	Portugal	24
United Kingdom	81	Denmark	16
Spain	60	Ireland	15
Netherlands	25	Luxembourg	6

European Council

The European Council comprises governmental representatives of the 12 member states, who are normally the heads of government and their foreign ministers. The president of the CEC is also a member.

The council is assisted by working parties and by a Permanent Representatives Committee (which comprises ambassadors of the member states).

The Commission of European Communities

The CEC has 17 members appointed by agreement among the participating governments and who are elected for four-year terms. The CEC members are independent of the governments and the council. The CEC is answerable to the European Parliament.

Other Institutions

There is also a Court of Justice, which has thirteen judges appointed for six years by the governments. The Court of Auditors has twelve members appointed by the council.

The Economic and Social Committee comprises 189 members representing various sectors of economic and social life. The Committee assists the Council and the CEC and is consulted before decisions are taken on a large number of subjects.

Legislation

Under the Treaty of Rome, the council and the CEC issue

- regulations,
- directives,
- decisions,
- recommendations, and
- opinions.

Regulations are binding in their entirety on all member states. Directives are binding on the member states to which they are addressed, regarding the results to be achieved, but leave the form and method of achieving the result to the discretion of the member states. Decisions may be addressed to a government, an enterprise, or a private individual, and are binding in their entirety on those to whom they are addressed. Recommendations and opinions are not binding.

Activities in the Field of Standardization

The general program for the elimination of technical barriers to trade in industrial products was adopted by the council on 28 May 1969. The division responsible for such activities is Division III\A\1 of the Directorate General for Internal Market and Industrial Affairs.

Directives adopted by the council and the CEC as well as proposals for directives submitted for adoption are published in the Official Journal of the European Communities.

10.3 COMMISSION OF THE EUROPEAN COMMUNITIES (CEC)

Commission of the European Communities
200, rue de la Loi
B-1049 Brussels
Belgium

Telephone:	32 2 235 0575
Facsimile:	32 2 235 6828

The CEC has been set up with 22 Directorates General, the major ones being

DG III	Internal Market and Industrial Affairs;
DG IV	Competition;
DG VIII	Development;
DG XII	Science, Research and Development;
DG XIII	Telecommunications, Information Industries and Innovation.

The DG XIII has six Directorates within it, these being

Directorate A Information Technology (Esprit);

Directorate B Information Industry and Market;

Directorate C Exploitation of Research and Technology Development, Technology Transfer and Innovation;

Directorate D Telecommunications;

Directorate E Support for Sector Activities; and

Directorate F Policy.

Directorate D — Telecommunications has an adviser with special responsibility for strategic aspects, and concentrates on

- telecommunication policy;
- analysis and exploratory studies;
- land-based and satellite infrastructure, audio-visual techniques, and COST; and
- data communications (with special responsibility for electronic data transfer projects including INSIS and CADDIA).

Directorate E's activities include

- economic, financial, and international aspects specific to telecommunication, the information industry, and innovation; and
- standards and type approval in the field of information technology and telecommunication.

In formulating EC policy, the CEC is assisted by committees, which consider the need to start work on standards through putting standards to practical use. These committees include

- Senior Officials Group for Information Technology Standardization (SOG-ITS); and
- Senior Officials Group for Telecommunications (SOG-T).

SOG-T set up a subsidiary Group for Analysis and Forecasting (GAP), which looked at the question of pan-European mobile communication.

These committees guide the CEC in policy matters, and have specific responsibility for assisting in implementation of the legislation in this area. The Public Procurement Sub-Committee in the Information Technology sector (PPSC-IT), while having wider interests than standardization, has decided that standards are one of its priority subjects.

10.4 EUROPEAN TELECOMMUNICATIONS STANDARDS INSTITUTE (ETSI)

European Telecommunications Standards Institute
B.P. 152
F-06561 Valbonne Cedex
France

Telephone:	33 929 44200
Facsimile:	33 936 54716

History of ETSI

At the beginning of the 1980s what became clear within the CEPT was that changes, increasing pressure, and accelerating development in telecommunication showed a need for closer and more balanced cooperation with other interested parties such as industry and users. In 1987, in the "Green Paper," the Commission of the European Communities, stated that, if the

European single market was to be a reality by the end of 1992, telecommunication products must flow freely across the borders and standardization needed to be realized.

ETSI was set up with a view to increasing the speed of standardization, replacing the principle of consensus with one of weighted voting, and permitting participants in addition to the PTTs (i.e., industry, public network operators, users, private service providers, and some research bodies) in the standardization process. The headquarters of the European Telecommunications Standards Institute (ETSI) are at Sophia Antipolis, near Nice in the south of France. ETSI has been set up to participate in telecommunication standardization in cooperation with CEN and CENELEC (European Standardization Organizations) and in cooperation with EBU (European Broadcasting Union).

ETSI was officially formed by the administrations of France and the United Kingdom, jointly depositing the statutes of the institute as an association governed by the French law of 1 July 1901. The inaugural General Assembly meeting took place on 30 March 1988 and the first Technical Assembly meeting was held on 1 June 1988. One of the main topics at the third General Assembly meeting on 19 July 1988 in Nice was Article 14 of the Rules of Procedure, which governs the voting procedure for the adoption of standards.

At the second Technical Assembly meeting on 25 October 1988, a budgeted work program for 1989 was established, together with the terms of reference for the project teams and CEPT working groups, which were to be transferred to ETSI.

In 1989, the re-examination and modification of ETSI's Rules of Procedure with regard to the adoption of standards enabled recognition of ETSI as "European Standardization Organization" in the field of telecommunication.

An agreement between the EEC Commission and ETSI regarding the standardization work of ETSI was negotiated and signed in 1989.

Organization

ETSI has a General Assembly, which is its governing body and determines ETSI's policy, appoints the director and deputy director, adopts the budget, and approves the audited accounts. Each member votes directly on all financial matters, but other decisions are taken on a weighted national vote.

The Technical Assembly is the highest authority within ETSI for the production and approval of technical standards, advises on the work to be undertaken, and indicates priorities. A number of technical committees and project teams answer to the Technical Assembly and address specific topics and issues. Each technical committee addresses a defined specialized area and comprises Europe's senior experts in that speciality. The project teams are full-time groups of experts carrying out studies and preparing draft standards in accordance with the terms of reference established by, and for consideration by, technical committees. The technical committees are discussed below.

The director assisted by the deputy director is responsible for assigning resources, preparing work schedules and priorities, and liaising with external bodies within the guidelines laid down by the General Assembly and decisions taken by the Technical Assembly.

There is a small secretariat for daily management of the institute. Technical committees, however, provide their own secretariat resources, usually from the organization of their chairs. In 1989, under the Technical Assembly, the ISDN Standards Management Group (ISM) was set up as were two committees, the Strategic Review Committee (SRC) and Intellectual Property Rights Committee (IPRC).

Membership

Members include

- national administrations,
- public network operators,
- manufacturers,
- users and private providers offering services to the public, and
- research bodies.

Any of the above who demonstrate an interest in European telecommunication standardization may join. Members of ETSI must be based in one or more of the CEPT countries.

Members may participate individually or be grouped on a national or European basis; for example, trade associations may be particularly keen on representing the interests of small to medium sized enterprises. National standards bodies may also be members.

Manufacturing companies can participate as parent companies, subsidiaries, or both. Already there are approximately 180 members representing the leading European telecommunication interests. Instead of full membership, observer status may be obtained, giving the participant the right to speak but not to vote.

Representatives from the EEC and EFTA have special status in ETSI as counsellors. Observers from Europe may participate with the right to speak but not the right to vote in the General Assembly and Technical Assembly. With the agreement of the Technical Assembly, observers may also participate in the technical committees. Observers from outside Europe may be invited by the Technical Assembly to participate in its meetings and those of the technical committees where reciprocal possibilities exist for ETSI members to attend equivalent meetings in the invitee's country.

Members pay an annual fee comprising one or more units of contribution based on either the turnover of the organization or, in the case of administrations, the Gross Domestic Product of the country. The value of a single unit is set each year by the projected net cost of the following year's work program. In addition, administrations contribute on a so-called "CEPT scale" to the overhead of the institute. Whether administrations pass this cost on to their members is a national matter. The budget for 1989 was a total of 5,910 ECUs.

Standards

Standards prepared by an appropriate working structure and approved by the ETSI Technical Assembly will be known as European Telecommunication Standards (ETS), with voluntary status. In some cases, the standards approved will be named Interim European Telecommunication Standards (I-ETS). This designation will be adopted whenever a standard requires a trial period or represents only a provisional solution.

Adoption of an ETS or an I-ETS follows a public enquiry and weighted national voting. The public enquiry is for recognized national standards organizations that have the exclusive responsibility for establishing their national position for voting on the relevant standard. In some cases, European telecommunication standards may be transferred into legal mandatory standards through procedures or legal instruments that are enforced by other authorities in the EEC. ETSI may also carry out prestandardization studies, with a view to guiding early developments and preparing the groundwork for future standards.

The technical committees are

- Technical Committee NA (Network Aspects);
- Technical Committee BT (Business Telecommunications);
- Technical Committee SPS (Signaling Protocols and Switching);
- Technical Committee TM (Transmission and Multiplexing);
- Technical Committee TE (Terminal Equipment);
- Technical Committee EE (Equipment Engineering);
- Technical Committee RES (Radio Equipment and Systems);
- Technical Committee GSM (Special Mobile Group);
- Technical Committee PS (Paging Systems);
- Technical Committee SES (Satellite Earth Stations);
- Technical Committee ATM (Advanced Testing Methods); and
- Technical Committee HF (Human Factors).

Most of these technical committees were transferred from CEPT, with the main difference being that the meetings of technical committees and sub-technical committees are open to the experts of all ETSI members. Reorganization of the technical committees and the addition of new ones was accomplished in 1989.

There are 35 sub-technical committees. To enable participation by all members, ETSI established a so-called "logging-in system," which means that engineers employed by the members can register their interest in particular technical committees and sub-technical committees. The engineers then receive the essential papers issued by the relevant technical committee or sub-technical committee. More than 1,000 engineers have been involved in standardization work by this means. At the end of 1988, the technical committees had produced 17 draft ETS for formal adoption by the Technical Assembly.

The program approved in the first meeting of the Technical Assembly (June 1988) foresaw the involvement of a total of 240 person-months in nine Project Teams. In 1988, Project Team 1 (ISDN Basic Access) and Project Team 4 (Voice Band Modems) completed their work.

10.5 CONFERENCE OF EUROPEAN POSTS AND TELECOMMUNICATIONS ADMINISTRATIONS (CEPT)

Liaison Office CEPT
Sellerstrasse 22
Case Postale 1283
CH-3001 Berne
Switzerland

Telephone: 44 31 622 081
Facsimile: 44 31 622 078

The Conférence Européenne des Administrations des Postes et des Télécommunications (CEPT) brings together the PTTs of 26 European countries from Iceland to Turkey and Portugal to Finland, including Yugoslavia, but excluding other Eastern European countries. CEPT, which is independent of any political or economic organization, was created at Montreux, Switzerland, in 1959 by PTTs aware of the necessity of tightening and institutionalizing their links. The aim was to harmonize and improve postal and telecommunication coordination and cooperation among European countries to form a homogeneous, coherent, and efficient unit on a continental scale.

Structure and Organization

The structure and organization of CEPT is set out in Table 10.1.

CEPT meets in Plenary Assembly every two or three years to consider questions affecting both posts and telecommunication to review the work of each commission. Between Plenary Assemblies, the Postal and Telecommunications Commissions maintain the continuity of work in their respective fields, each being empowered to create specialized committees or working parties, and act on technical decisions, contacts with other bodies, and its own organisation. Overall continuity is provided by the CEPT Liaison Office, based in Berne and the Managing Administration for the next term. This role is currently filled by the United Kingdom, with the Department of Trade and Industry acting as chair and secretary of CEPT, and assuming responsibility for organizing the next Plenary Assembly in London in 1990. The privatized former British Post Office, now British Telecom, is responsible for chairing the two commissions.

Telecommunication

The Telecommunications Commission ("T" Com) is the body responsible

Table 10.1. CEPT Structure and Organization

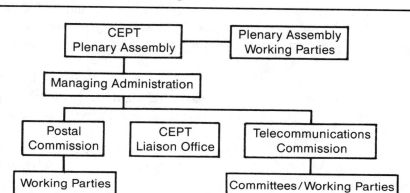

for determining telecommunication policy within CEPT. The aim is to achieve the effective operation and interconnection of international public telephone services among member countries, to promote the development of new services and products, and to coordinate, as appropriate, CEPT's response to developments in international telecommunication, notably within the ITU.

"T" Com has nominated five specialized committees to coordinate activities across the entire range of transmission media (transoceanic, satellite, radio, terrestrial) and also to consider both the technical and commercial implications of service developments. Each committee has established a number of working groups and, in some cases, specialized project teams. Business services by satellite, telemarketing, X.25, pan-European digital networks, and RACE are all featured in "T" Com's current activity.

Traditionally, "T" Com was empowered to draft and approve technical standards for services and equipment. In 1988, however, CEPT responded to European Commission proposals recommending the liberalization of the European telecommunication market by laying the foundations for the independent European Telecommunications Standards Institute (ETSI), whose membership includes not only CEPT members, but also representatives from industry and user groups. This development, together with the trend toward the separation of regulatory and operational functions, the emergence of competition in the provision of telecommunication services, and the participation of international bodies and national governments in developing telecommunication policy, means that "T" Com must continually adapt. An increasingly important aspect of its work as a result of these trends is to develop closer relations with the European Commission and EFTA. In this way, "T" Com is able both to respond to and anticipate communication developments and requirements, and to make its voice heard in political circles.

The flexibility of this forum, which attracts the most senior CEPT policy-makers to its sessions, means that "T" Com plans to continue to play a significant role in ensuring cooperation in the development and provision of European telecommunication services.

10.6 EUROPEAN COMMITTEE FOR STANDARDIZATION (CEN)

European Committee for Standardization
2, rue Brederode, Bte5
B-1000 Brussels
Belgium

Telephone:	32 2 519 6825
Facsimile:	32 2 219 6819

The European Committee for Standardization (CEN) is an association of the national standards bodies in the countries belonging to the European Community (EC) and the European Free Trade Association (EFTA), in addition to Spain (totaling 16 members).

Some 122 European standards (ES, or EN for *norme*) and Harmonization Documents (HD), 194 draft European standards (prEN) and draft Harmonization Documents (prHD), and 3 CEN Reports (CR) have been adopted or prepared.

The standards bodies of the European countries, the CEN and CEC, have set up a cooperative agreement with a view to establishing a correspondence table among the national standards of the countries involved.

10.7 EUROPEAN COMMITTEE FOR ELECTROTECHNICAL STANDARDIZATION (CENELEC)

CENELEC
2, rue Brederode, Bte5
B-1000 Brussels
Belgium

| Telephone: | 32 2 519 6871 |
| Facsimile: | 32 2 519 6819 |

The European Committee for Electrotechnical Standardization (CENELEC), a nonprofit technical organization set up under Belgian law, is composed of the National Electrotechnical Committees of 18 countries in Western Europe, most of which (except for Iceland and Luxembourg) are also members of the IEC.

Organization

CENELEC has a General Assembly (AG), a Technical Board (BT), technical committees, working groups, and three programming committees:

- CPC 1 — basic electrotechnology;
- CPC 2 — electrical power equipment; and
- CPC 3 — electrical consumer goods and control equipment.

The CPCs make recommendations to the GA, with follow-up being normally entrusted to the TB. The work of 54 technical committees and subcommittees is controlled by the TB.

Standstill Agreement

Once CENELEC has selected an international standard or any other useful document as a reference and submits it to the agreed procedure, all National Electrotechnical Committees must stop any national work on the subject. The members are not to publish during a given period a new or revised national standard that is not completely in line with a CENELEC standard in existence or under preparation. Members are not to take any action that may prejudice the intended harmonization. The same procedure applies when the AG or BT decides to initiate new work on a specific item within a technical body of CENELEC.

Transposition

After a formal vote of approval the standards are ratified by the CENELEC TB. All National Electrotechnical Committees are then obliged to implement the CENELEC standards.

Uniform implementation by each CENELEC member is required in compliance with the CEN-CENELEC common rules for standards work.

Marks Committee

The CENELEC Marks Committee deals with problems of mutual recognition of national *marks of conformity* for electrical products and general questions concerning specific certification schemes.

The CCA and HAR Agreements

The CENELEC Certification Agreement (CCA) on mutual recognition of test results for approval of electrical equipment operates within the framework of the CENELEC Marks Committee. The CENELEC HAR certification arrangement for low-voltage cables and cords, also within the framework of the CENELEC Marks Committee, organizes certification on the basis of harmonized standards. The manufacturer obtains a licence to mark its cables or cords with the national mark of conformity of the licensing body together with the marking "HAR." Cables and cords bearing this marking along with any national marks are recognized in all member countries as being in conformity with the relevant CENELEC standards.

The CECC System for Certification

The CENELEC Electronic Components Committee (CECC) deals with standards for and quality assessment of electronic components.

The CECC system harmonizes specifications and quality assessment procedures for electronic components, and has approved an internationally recognized *mark or certificate of conformity*. The quality assessment procedures used are either *qualification approval* for components that meet requirements given in a detailed specification or *capability approval,* where technology which has precisely defined boundaries is approved. Implementation of the system's rules is the responsibility of member countries, each represented by a National Authorized Institution (ONH).

Inspection and surveillance of the assessment procedure is ensured by the independent Electronic Components Quality Assurance Committee (ECQAC). There is also a scheme for certification of information technology equipment (ECITC).

CENCER Steering Committee

Any activity in the field of CEN certification is denoted as CENCER. CEN certification activities are managed by the CENCER Steering Committee. CENCER offers two alternatives. Its primary objective is the operation of a conformity marking system, which provides schemes for the application of the CEN conformity-marked products manufactured in accordance with the requirements of a European standard or HD. This system is based on initial type testing and continuing quality control, including sample testing of the products. CENCER is a third-party certification system, requiring fully harmonized standards. The alternative is the CENCER system for mutual recognition of test and inspection results based on standards. It is a far more flexible system. The latter is based on the reciprocal recognition by CEN members of each other's test and inspection results and applicable to standards that are not fully harmonized. This system, however, does not necessarily lead to the granting of the CEN conformity mark.

CENELEC has a Marks Committee that deal with problems of mutual recognition of national marks of conformity, certificates of conformity, and other means of proving compliance with standards. There is a special scheme for low-voltage cables and cords known as the HAR harmonized scheme. A special committee, the CENELEC Electronic Components Committee (CECC), deals with the highly specialized area of quality assessment for electronic components. CEN has one certification mark. CENELEC has two collective marks, one for cables and cords, and the CECC mark for electronic components.

The main areas of CENELEC's activities in European electrotechnical standardization are illustrated in Table 10.2.

10.8 THE TELECOMMUNICATION STANDARDIZATION PROCESS IN THE EUROPEAN ECONOMIC COMMUNITY

The Common Market

The EEC is not a standardization organization as such. It has, however, a direct interest in and a great effect on standardization. The EEC perceives standards as one source of technical barriers and believes that standardization can be a powerful tool in the creation of a true common market. The EEC recognizes that technical barriers created by different national product regulations and standards not only add extra costs, but also distort production patterns by raising inventory holding costs, discouraging business cooperation, and preventing corporations from benefiting from the economies of scale that a larger, unified European market can offer. The harmonization approach has therefore become the cornerstone of the EEC's activities since 1965.

The expectation is that by 1992 there will no longer exist any physical,

Table 10.2. CENELEC and European Electrotechnical Standardization

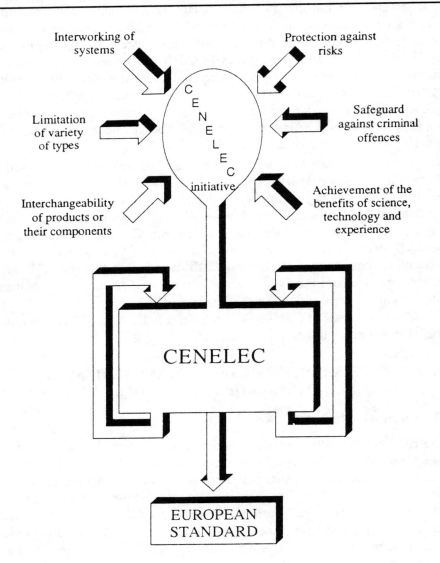

fiscal, or technical barriers among EEC member countries. By that time, if an American company exports to one EEC country, that firm can export to all EEC countries.

In information technology and telecommunication the CEC wanted much greater precision and more rapid decision making to ensure compatibility, intercommunication, and interworking among the users and operators throughout the EEC.

Barriers and Standstill

To prevent the creation of new technical barriers, Directive 83/189/EEC obliges member states to notify the CEC in advance of all draft regulations and standards concerning technical specifications that members intend to introduce within their own territories.

The EEC is empowered to call a temporary halt to any national standards activity that may create new barriers within the community and to substitute activity on a European basis.

Under this "standstill" agreement, member states must refrain from adopting a national standard for a limited period, during which a European Standard can be agreed; if not, the standstill comes to an end.

Barriers caused by differing technical rules and approval procedures for devices for attachment to networks are being removed as a result of Directive 86/361/EEC, which provides for the development of common technical specifications ("NETs").

Barriers to information flow caused by network incompatibilities are tackled by one of the articles in Decision 87/95/EEC. Public networks must offer a service such that end-to-end compatibility is provided for systems conforming to the functional standards covering network access.

Separation of Legislation

Directives incorporating detailed technical specifications resulted in long delays because of the unanimity requirement in EC decision making and in the procedures established for the implementation of Article 100 of the Treaty of Rome. A new approach to standardization was undertaken in 1986, designed to abandon gradually the practice of including technical documents in the directives and replacing these with references to European standards or, in exceptional cases, national standards. Instead of preparing individual EEC directives aimed at removing isolated problems connected with identified barriers to trade, the CEC will prepare directives to cover technical areas required for the operation of a unified internal market. The principle of reference to European standards prepared by CEN, CENELEC, or ETSI for all detailed technical requirements is a central part of this fundamental policy, which has been endorsed by the European Free Trade Association (EFTA).

Under the "new approach" technical and legal aspects are separated. The legal aspects and basic goals are defined in instruments such as directives and decisions, and the technical aspects to achieve those goals are defined in standards. The role of the directives is to establish the level of basic requirements to be observed, and not to specify the details of methods or to publish technical specifications. The process comprises four fundamental principles:

- development of directives concerning safety and all other public interest aspects that contain only basic requirements;
- reference to standards as proof of conformity to the basic requirements of the directives;
- no regulatory obligation to conform to the standards, meaning that any method of proof may be employed; and
- assumption of conformity to basic requirements of directives when conformity with a harmonized standard exists.

Programs to Encourage Technology Standards

The CEC may also initiate or encourage standardization bodies in the production of standards through "mandates" (*Bons de Commande*) entrusted to the standardization bodies under Directive 83/189/EEC and within the scope of a "framework contract," concluded in 1985. In such cases, the CEC bears a fair proportion of the costs of producing the standards. Standardization bodies nevertheless retain the right to decide whether to accept the mandate, to determine the technical specification, and ultimately to vote to approve specification as a standard or not.

About 70 mandates have been entrusted to CEN-CENELEC with the collaboration of CEPT. In addition to these formal mechanisms of legislation and the standardization machinery, the CEC can intervene by initiating projects and funding or subsidizing work that leads to more effective standardization. This work can be "upstream" in the formal standardization process, for example, in supporting studies aimed at establishing future requirements for standards; "side-stream," in which case help is given during the process; or "downstream," in which case measures are taken to ensure the effective implementation of agreed standards.

The most important downstream activity for which about 20 million ECUs of funding have been provided is the establishment of conformance testing services (CTS 1 and 2).

The standardization policy also has its application in the management of research and development programs funded by the EC in the information technology and telecommunication fields, such as ESPRIT, RACE, or projects in other areas, which include information technology. Some examples are

- the reference to existing standards when applicable in procurement of equipment or execution of the research work;
- the link to be established with the standardization bodies; and
- transfer of results as a contribution to the standardization process.

In some cases, the projects can act as a proving ground for new standards, and the experience can be channeled into the standardization arena. Indeed, an early application of Open Systems Interconnection (OSI) standards in the ESPRIT program (for communication between participants) demonstrated that the basic OSI standards could not be used without the added value provided by functional standards. This experience created one of the driving forces for the European Functional Standards program.

The CEC also promotes the use of European standards, in particular those of the ISO, in programs that concern

- public users, as in the INSIS or CADDIA programs;
- private users, as in the TEDIS program, related to trade data interchange between manufacturers or service suppliers.

Other Europe-wide projects backed by the CEC, such as RARE and COSINE, aim at the rapid establishment of an environment in which data communication services can be made available to users in academic and industrial research organizations through common standards.

Joint Information Technology Steering Committee (ITSTC)

The Joint Information Technology Steering Committee (ITSTC) comprises CEN, CENELEC, and ETSI.

ITSTC programs the preparation of European standards for the introduction in Europe of the OSI model developed by ISO in the information technology field. ITSTC, supported in this initiative by the CEC, EFTA, the major users, and European manufacturers, defined the concept of a "functional" standard, which is essential for effective application of the OSI model and has rapidly developed initial applications.

European Standard (EN)

Publications resulting from technical work of CEN-CENELEC and issued for implementation as identical national standards are designated as European Standards (ENs).

Harmonization Document (HD)

Publications resulting from technical work of CENELEC and issued for implementation of their technical content at national level are designated Harmonization Documents (HDs). HDs are established if transposition into identical national standards is believed unnecessary or impractical. The HDs must be implemented at a national level, either by the issuance of a corresponding national standard or, as a minimum, by the public announcement of the HD number and title. In any case, no conflicting national standards may continue to exist after the date fixed by the TB for the withdrawal of such standards. The absence of a corresponding national standard is not an obstacle to harmonization.

European Prestandard (ENV)

ENVs may be established as prospective standards for a provisional application in all technical fields where the innovation rate is high or there is an urgent need for technical guidance.

ENVs are usually prepared by a CENELEC technical body, which will adopt the document at a special voting meeting or through a written voting procedure. Any conflicting national standards may be kept in force until the ENV is converted into an EN.

The lifetime of these prestandards is normally limited to three years, but the TB can convert the ENV into an EN after a formal vote, extend its lifetime, replace it by a revised ENV, or withdraw it.

Technical Recommendations (NETS)

Technical Recommendations (called *Normes Européennes de Télécommunications*) are given mandatory force according to the legal procedures established in Directive 86/361/EEC. NETs are common technical specificaions covering access to networks and equipment attachment to networks and are normally used by CEPT. In 17 countries which signed a Memorandum of Understanding (MOU) in November 1985 (including all the EEC and five EFTA countries) there is a commitment by the signatories to refer to NETs in legislation governing type approval for attached equipment. NETs are now issued by ETSI. An illustration of the first functional standards issued and the first working program of NETs are in Tables 10.3 and 10.4, respectively.

European Telecommunication Standards (I-ETS and ETS)

ETSI may issue voluntary Interim European Telecommunication Standards (I-ETS) and European Telecommunication Standards (ETS). These are described more fully under ETSI (Section 10.4) above.

The Joint European Standardization Organization (CEN-CENELEC)

Since CEN and CENELEC have agreed to adopt the same internal bylaws, common practice has become to combine the two organizations under the name "Joint European Standardization Organization." One the goals of CEN and CENELEC is the harmonization of European standards, *harmonization* being defined as

- existing when equipment manufactured in one country automatically complies with the requirements of the national standards of the other member countries. Such equipment can then be sold in the market of each of the members.

CEN and CENELEC are international associations incorporated under Belgian law. CEN and CENELEC unite the European members of the ISO and the IEC, respectively. CEN members (16 of the 90 ISO member countries) perform approximately two-thirds of all technical work in ISO and provide financial support of similar magnitude.

There are two major types of documents produced by CEN and CENELEC (described in Section 10.6 above), namely,

- the European Standard (ES), and
- a Harmonization Document (HD).

An ES acquires the status of a European standard if it is developed by

Table 10.3. The First Functional Standards (from Commission of the European Communities, DG XIII, May 1988)

The first functional standards issued

ENV 41 101	LAN — Provision of the OSI connection-mode transport service on a CSMA/CD single LAN (Connection service multiple access — collision detection)
ENV 41 102	LAN — Provison of the OSI connection-mode transport service on a CSMA/CD LAN in a single or multiple LAN configuration
ENV 41 103	LAN — Provision of the OSI connection-mode transport service in an end system on a CSMA/CD LAN
ENV 41 104	Packet-Switched Data Networks — permanent access: OSI connection-mode transport service over either the OSI connection-mode network service or the T70 Case
ENV 41 105	Packet-Switched Data Network (PSDN) — switched access
ENV 41 106	Digital Data Circuit (CSDN): Provision of the OSI connection-mode transport and the OSI connection-mode network service (CO-NS)
ENV 41 107	CSDN-CO-NS Circuit-switched data network, connection oriented network service — permanent and switched access
ENV 41 109	Information systems interconnection: Local area networks provision of the OSI connection-mode transport service using connectionless-mode network service in an end system on a token-ring single LAN
ENV 41 110	Information systems interconnection: Local area network provision of the OSI connection-mode transport service and the OSI connection-less-mode network service in an end system on a token-ring LAN in a single or multiple configuration
ENV 41 201	Private message handling system (MHS)
ENV 41 202	Public message handling system (Administration management domain)
ENV 501	Graphic character repertoire and coding for interworking with CEPT videotex services
ENV 502	Graphic characters repertoire and coding for interworking with CEPT teletex services
ENV 41 503	European graphic character repertoires and coding
ENV 901	'Triple X' between character-mode and packet-mode terminals via a PAD

Note: ENV 41 001 (terminal connector) in an ENV but *not* a functional standard.

Table 10.4. The First Working Program of NETs (from Commission of the European Communities, DG XIII, May 1988)

The first working programme of NETs	
NET 1	X.21 access
NET 2	X.25 access
NET 3	ISDN basic access
NET 4	PSTN basic access
NET 5	Digital telephony
NET 6	Analogue modem (V.32)
NET 7	Group 3 facsimile
NET 8	Teletex
NET 9	ISDN terminal adaptor (X.21)
NET 10	Pan-European digital cellular radio access

CEN or CENELEC and therefore rests on a consensus of the majority. An HD differs from a European standard in that it records technical differences or legislative obstacles, and its primary purpose is to guarantee uniform application of international standards.

The French had worked with experimental standards for some years, and in 1985 CEN-CENELEC decided to use ENVs to allow European standards to reflect technological developments as rapidly as possible. Existing national standards that conflict with an ENV may be maintained until it becomes a European standard. In principle, the interim standard is valid for three years. This period may be extended to five years. By mid-1986, there were three ENVs published, two dealing with local area networks (interconnection of systems) and a third on private electronic mailbox systems.

CEN-CENELEC has also been asked to integrate into European standardization those draft standards developed outside their organization. Both CEN and CENELEC are empowered, when necessary, to create joint technical working committees and working groups.

CENELEC has chosen to take IEC standards and to convert these into harmonization documents (HDs or ESs). CEN has decided to undertake its work only at the European level in fields not covered by international standardization, where a specific need has been identified.

Organization of the Joint European Standardization Organization

CEN and CENELEC are composed of a General Assembly, a General Council, a Technical Office, technical committees, working groups, certification committees, and a central secretariat.

CEN and CENELEC deal only with specific, limited objectives for which a need for standards has been clearly established. If the subject has been dealt with by ISO or IEC in an appropriate fashion, CEN and CENELEC tailor their activities to ensure that the work performed by these organizations is applied and complemented.

During the General Assemblies of CENELEC, on September 27, 1985, and CEN, October 3, 1985, five sets of joint rules were adopted for preparation of a European Standard and a Harmonization Document. These rules, which became effective in January 1986, are

- balloting on implementation,
- *status quo* agreements,
- development of draft European Standards,
- association of CEN-CENELEC with other organizations involved in European standardization, and
- joint technical CEN-CENELEC projects.

Balloting on implementation generally works on the simple majority of ballots cast. In four exceptional cases, however, the ballots cast are weighted, for example, on

- final approval of European standards and HDs,
- final approval of interim standards,
- the beginning and end of *status quo,* and
- approval of differences.

For adoption of a proposal based on a weighted ballot, a number of conditions related to the number of members voting against and in favor must be met. Once an ES or HD has been approved, all members are required to implement it.

Members inform the CEN-CENELEC secretariat of their annual standardization program and transmit the draft text. Then, the secretariat retransmits this information to the CEC, the EFTA secretariat, and the standardization organization of each member country.

From the time that a decision to develop an ES or HD has been made until the document is prepared, CEN members must refrain from publishing final versions of corresponding national standards. This is known as the *status quo* or standstill agreement.

The ICON Project

The ICON project, a comparative index of European standards, is being developed with EEC financial assistance. This database contains standards in effect in the member countries. Initially, this computerized index will contain the references of only those national standards which produce the contents of international standards (ISO, IEC) or European standards, or those which deal with the same subject. This index will enable the user, by using the reference of a national standard, to find the reference of an identical standard of another country and to know in what language it is published. The EEC has granted a mandate to CEN-CENELEC to establish the database for comparisons among the national standards. Comparisons are made among ISO, IEC, CEN, and CENELEC standards that will constitute the key points connecting the national standards of the different countries. ICON will allow for comparisons by subject, but not content, among standards at the three levels proposed by the ISO guideline: identity, equivalence, and non-equivalence. Where there is no key European or international point, the standards are compared by commercial sectors.

10.9 IMPLEMENTATION OF THE GREEN PAPER

The emerging service economy increasingly depends on telecommunication as its basic infrastructure, providing the means to transport and trade a range of other services. The economic benefits for Europe in 1992 will depend in large measure on completing the internal market for telecommunication. The cost of "non-Europe" was high in telecommunication as by the early 1980s Europe was lagging in

- information technology, and particularly integrated circuit components; and
- deregulation of telecommunication.

The CEC's June 1987 "Green Paper" (Towards a Dynamic Economy — Green Paper on the Development of the Common Market for Telecommunications Services and Equipment, COM(87) 290, Brussels, 30 June 1987) sought to overcome these problems by proposing a program of regulatory change and technological development for the EEC.

The CEC has achieved rapid agreement by the EC on

- future development of telecommunication in the EEC and common infrastructure projects, in particular concerning the principal future stages of network development (ISDN), digital mobile communication, and the introduction of future broadband communication;
- creation of a community-wide market for terminals and equipment, and in particular the promotion of open standards throughout Europe to give equal opportunity to all market participants;
- launch of a program of precompetitive and "prenormative" research and development covering the technologies required for integrated broadband communication (the RACE program);
- promotion of the introduction and development of advanced services and networks in the less-favored peripheral regions of the EEC; and
- building common European positions with regard to international discussions in telecommunication.

Table10.5 shows community decisions taken in the field of telecommunication since 1984, including EC decisions, resolutions, recommendations, regulations, and directives.

Table 10.6 summarizes the Green Paper, in particular how it will involve implementation of lines of action defined by the EC in December 1984 and six new lines of action:

- ensuring the long-term convergence and integrity of the network infrastructure via the RACE program, proposals for the introduction of digital mobile communication, coordinated introduction of ISDN, and the STAR program for advancing infrastructure in the region to increase economic cohesion;
- full mutual recognition of type approval for terminal equipment;
- opening access to public telecommunication procurement contracts;
- creation of the European Telecommunications Standards Institute (ETSI);

Table 10.5. EEC Decisions in Telecommunication Since 1984

COUNCIL RECOMMENDATION OF 12 NOVEMBER 1984 concerning the implementation of harmonization in the field of telecommunications (84/549/EEC; O.J. L298/49)

COUNCIL RECOMMENDATION OF 12 NOVEMBER 1984 concerning the first phase of opening up access to public telecommunications contracts (84/550/EEC; O.J. L298/51)

COUNCIL DECISION OF 25 JULY 1985 on a definition phase for a Community action in the field of telecommunications technologies - R & D programme in advanced communications technologies for Europe (RACE) (85/372/EEC; O.J. L210/24)

COUNCIL RESOLUTION OF 9 JUNE 1986 on the use of videoconference and videophone techniques for intergovernmental applications (86/C 160/01; O.J. C160/01))

COUNCIL DIRECTIVE OF 24 JULY 1986 on the initial stage of the mutual recognition of type approval for telecommunications terminal equipment (86/361/EEC; O.J. L217/21)

COUNCIL REGULATION OF 27 OCTOBER 1986 instituting a Community programme for the development of certain less-favoured regions of the Community by improving access to advanced telecommunications services (STAR programme) (86/3300/EEC; O.J. L305/1)

COUNCIL DIRECTIVE OF 3 NOVEMBER 1986 on the adoption of common technical specifications of the MAC/packet family of standards for direct satellite television broadcasting (86/529/EEC; O.J. L311/28)

COUNCIL DECISION OF 22 DECEMBER 1986 on standardisation in the field of information technology and telecommunications (87/95/EEC; O.J. L36/31)

COUNCIL RECOMMENDATION OF 22 DECEMBER 1986 on the coordinated introduction of the Integrated Services Digital Network (ISDN) in the European Community (86/659/EEC; O.J. L382/36)

COUNCIL RECOMMENDATION OF 25 JUNE 1987 on the coordinated introduction of public pan-European cellular digital land-based mobile communications in the Community (87/371/EEC; O.J. L196/81) and COUNCIL DIRECTIVE on the frequency bands to be reserved for the coordinated introduction of public pan-European cellular digital land-based mobile communications in the European Community (87/372/EEC; O.J. L196/85)

COUNCIL DECISION OF 5 OCTOBER 1987 introducing a communications network Community programme on trade electronic data interchange systems (TEDIS) (87/499/EEC; O.J. L285/35)

COUNCIL DECISION OF 14 DECEMBER 1987 on a Community programme in the field of telecommunications technologies - research and development (R&D) in advanced communications technologies in Europe (RACE programme) (88/28/EEC; O.J. L16/35)

Table 10.5. EEC Decisions in Telecommunication Since 1984 (Continued)

In the framework of the strategy as expanded by the Green Paper, the following decisions and proposals have been made since the start of 1988

COMMISSION DIRECTIVE OF 16 MAY 1988 on competition in the markets in telecommunications terminal equipment (88/301/EEC; O.J. L131/73)

COUNCIL RESOLUTION OF 30 JUNE 1988 on the development of the common market for telecommunications services and equipment up to 1992 (88/C 257/01; O.J. C257/1)

PROPOSAL FOR A COUNCIL DIRECTIVE OF 11 OCTOBER 1988 on the procurement procedures of entities operating in the telecommunications sector (COM(88)378)

PROPOSAL FOR A COUNCIL DIRECTIVE OF 9 JANUARY 1989 on the establishment of the internal market for telecommunications services through the implementation of Open Network Provision (COM(88)825)

COUNCIL RESOLUTION OF 27 APRIL 1989 concerning the strengthening of the further coordination of the introduction of the integrated services digital network (ISDN) in the Community up to 1992

COUNCIL RESOLUTION OF 27 APRIL 1989 concerning standardization in the fields of information technology and telecommunications

PROPOSAL FOR A COUNCIL RECOMMENDATION of 5th June 1989 on the coordinated introduction of pan-European land-based public radio paging in the Community (COM(89)166) and
PROPOSAL FOR A COUNCIL DIRECTIVE of 5th June 1989 on the frequency bands to be reserved for the coordinated introduction of pan-European land-based public radio paging in the Commumnity (COM(89)166)

PROPOSAL FOR A COUNCIL DIRECTIVE of 27th July 1989 on the approximation on the laws of the Member States concerning telecommunications terminal equipment, including the mutual recognition of their conformity (COM(89)289)

REVISED PROPOSAL FOR A COUNCIL DIRECTIVE of 10th August 1989 on the establishment of the Internal Market for telecommunications services through the implementation of Open Network Provision (O N P) (COM(89)325)

This list does not include Community decisions in closely related fields, in particular high definition television (Council Decision of 27 April 1989), the IT application programmes (Drive: 88/416/EEC; O.J. L206/1; Delta: 88/417/EEC, O.J. L206/20; Aim: 88/577/EEC; O.J. L314/22); the implementation of the information services market (88/524/EEC; O.J. L288/39); and the Insis and Caddia (85/214/EEC; O.J. L96/35 - 86/23/EEC; O.J. L33/28 - 87/288/EEC; O.J. L145/86) programmes.

Table 10.6. The Green Paper's Proposed Positions

The general objective of the positions set out is the development in the Community of a strong telecommunications infrastructure and of efficient services: providing the European user with a broad variety of telecommunications services on the most favourable terms, ensuring coherence of development between Member States, and creating an open competitive environment, taking full account of the dynamic technological developments underway.

A) *Acceptance of continued exclusive provision or special rights for the Telecommunications Administrations regarding provision and operation of the network infrastructure. Where a Member State chooses a more liberal regime, either for the whole or parts of the network, the short and long term integrity of the general network infrastructure should be safeguarded.*

   Closely monitored competitive offering of two-way satellite communications systems will need further analysis. It should be allowed on a case-by-case basis, where this is necessary to develop Europe-wide services and where impact on the financial viability of the main provider(s) is not substantial.

   Common understanding and definition regarding infrastructure provision should be worked out under E) below.

B) *Acceptance of continued exclusive provision or special rights for the Telecommunications Administrations regarding provision of a limited number of basic services, where exclusive provision is considered essential at this stage for safeguarding public service goals.*

   Exclusive provision must be narrowly construed and be subject to review within given time intervals, taking account of technological development and particularly the evolution towards a digital infrastructure. 'Reserved services' may not be defined so as to extend a Telecommunications Administration service monopoly in a way inconsistent with the Treaty. Currently, given general understanding in the Community, voice telephone service seems to be the only obvious candidate.

C) *Free (unrestricted) provision of all other services ('competitive services', including in particular 'value-added services') within Member States and between Member States (in competition with the Telecommunications Administrations) for own use, shared use, or provision to third parties, subject to the conditions for use of the network infrastructure to be defined under E).*

   'Competitive services' would comprise all services except basic services explicitly reserved for the Telecommunications Administrations (see B).

D) *Strict requirements regarding standards for the network infrastructure and services provided by the Telecommunications Administrations or service providers of comparable importance, in order to maintain or create Community-wide interoperability. These requirements must build in particular on Directives 83/189/EEC and 86/361/EEC, Decision 87/95/EEC and Recommendation 86/659/EEC.*

   Member States and the Community should ensure and promote provision by the Telecommunications Administrations of efficient Europe-wide and worldwide

212

Table 10.6. The Green Paper's Proposed Positions (Continued)

communications, in particular regarding those services (be they reserved or competitive) recommended for Community-wide provision, such as according to Recommendation 86/659/EEC.

E) Clear definition by Community Directive of general requirements imposed by Telecommunications Administrations on providers of competitive services for use of the network, including definitions regarding network infrastructure provision.

This must include clear interconnect and access obligations by Telecommunications Administrations for trans-frontier service providers in order to prevent Treaty infringements.

Consensus must be achieved on standards, frequencies, and tariff principles, in order to agree on the general conditions imposed for service provision on the competitive sector. Details of this Directive on Open Network Provision (O N P) should be prepared in consultation with the Member States, the Telecommunications Administrations and the other parties concerned, in the framework of the Senior Officials Group on Telecommunications (SOG-T).

F) Free (unrestricted) provision of terminal equipment within Member States and between Member States (in competition with Telecommunications Administrations), subject to type approval as compatible with Treaty obligations and existing Directives. Provision of the first (conventional) telephone set could be excluded from unrestricted provision on a temporary basis.

Receive Only Earth Stations (ROES) for satellite down-links should be assimilated with terminal equipment and be subject to type approval only.

G) Separation of regulatory and operational activities of Telecommunications Administrations. Regulatory activities concern in particular licensing, control of type approval and interface specifications, allocations of frequencies, and general surveillance of network usage conditions;

H) Strict continuous review of operational (commercial) activities of Telecommunications Administrations according to Articles 85, 86 and 90, EEC Treaty. This applies in particular to practices of cross-subsidization of activities in the competitive services sector and of activities in manufacturing;

I) Strict continuous review of all private providers in the newly opened sectors according to Articles 85 and 86, in order to avoid the abuse of dominant positions;

J) Full application of the Community's common commercial policy to telecommunications. Notification by Telecommunications Administrations under Regulation 17/62 of all arrangements between them or with Third Countries which may affect competition within the Community. Provision of information to the extent required for the Community, in order to build up a consistent Community position for GATT negotiations and relations with Third Countries.

- common definition of an agreed set of conditions for Open Network Provision (ONP) to service providers and users;
- common development of Europe-wide services;
- common definition of a coherent European position on the future development of satellite communication in the community;
- common definition of telecommunication services and equipment with regard to relations with non-EC countries; and
- common analysis of social impact.

Public consultation on the Green Paper's proposals has emphasized the need for

- the current and future integrity of the basic network infrastructure to be maintained or created; and
- the provision by PTTs of telecommunication services alongside other suppliers.

Therefore, while there may be complex problems of regulation, there will be a continuing strong role for PTTs, the provision of network infrastructure, and strong emphasis on Europe-wide standards. The financial viability of PTTs hence must be safeguarded to ensure a build-up of telecommunication infrastructure and investment.

Table 10.7 sets out the timetable for implementing the Green Paper, which was submitted to the EC, the European Parliament, and the EEC's Economic and Social Committee.

There has been significant progress to date in implementing the agreed timetable, which involves directives on

- opening of the terminal equipment market to competition;
- opening of the services market to competition;
- opening of receive-only antennas not connected to the public network;
- implementation of the general principle that tariffs should follow overall cost trends;
- separation of regulatory and operational activities;
- definition of ONP;
- establishment of the ETSI;

Table 10.7. Timetable for Implementation of the Green Paper

1   Rapid full opening of the terminal equipment market to competition by 31 December 1990 at the latest

2   Progressive opening of the telecommunications services market to competition from 1989 onwards, with all services other than voice, telex and data communications to be opened by 31 December 1989. This should concern in particular all value-added services. Special consideration should apply to telex and packet- and circuit-switched data services

3   Full opening of receive-only antennas as long as they are not connected to the public network, by 31 December 1989

4   Progressive implementation of the general principle that tariffs should follow overall cost-trends. A review of the situation achieved by 1 January 1992 was announced

5   Clear separation of regulatory and operational activities

6   Definition of Open Network Provision (ONP). This was initially to cover access to leased lines, public data networks, and ISDN. Directives to Council to be submitted according to progress of definition work

7   Establishment of the European Telecommunications Standards Institute

8   Full mutual recognition of type approval for terminal equipment

9   Introduction - where this does not yet apply - of value-added tax to telecommunications, by 1 January 1990 at the latest

10   Guidelines for the application of competition rules to the telecommunications sector, in order to ensure fair market conditions for all market participants

11   Opening of the procurement of Telecommunications Administrations.

- full mutual recognition of type approval for terminal equipment;
- introduction of a value-added tax on telecommunication services;
- guidelines for the application of competitive rules; and
- opening of procurement.

10.10 EUROPEAN COMPUTER MANUFACTURERS ASSOCIATION (ECMA)

> European Computer Manufacturers Association
> 114, rue du Rhône
> CH-1204 Geneva
> Switzerland
>
> Telephone: 41 22 735 36 34
> Facsimile: 41 22 786 52 31

The ECMA, a nonprofit association, was formed in May 1961 in an endeavor to participate actively in the standardization of basic digital computer operational techniques including information coding and telecommunication.

Forty-eight companies in either of two classes of membership actively participate in the ECMA Work. There are 32 ordinary members, who develop, manufacture, and market data-processing equipment in Europe. There are also 16 associate members, who have an interest in Europe on specific subjects of one or more of the ECMA technical committees.

The associate members have full voting rights in the technical committees. ECMA functions very much like a miniature ISO and has a structure of some fourteen technical committees (TCs) and eight task groups (TGs). The most interesting of these is TC32 (Communications, Networks and Systems Interconnection). Among its TGs are TG2 (Distributed Interactive Processing), TG4 (OSI Management), TG6 (Private Switching Networks), TG9 (Security), and TG11 (CSTA). ECMA produces its own standards on matters that it considers important and which either have not been or are being studied by the principal organizations. In the former case, the ECMA standard finds wide recognition. In the latter, the standard serves as a complete, final proposal supported by industry and often influences the ultimate outcome. Seventeen ECMA standards have been endorsed by ISO and IEC under their fast-track procedure and published as ISO-IEC international standards. Three ECMA standards have been processed by CENELEC as interim European standards (ENVs).

10.11 EUROPEAN CONFERENCE OF ASSOCIATIONS OF TELECOMMUNICATION INDUSTRIES (EUCATEL)

EUCATEL is an organization of European national trade associations of telecommunication equipment manufacturers (e.g., TEMA for the United Kingdom, Fabrimétal for Belgium, *et cetera*). EUCATEL is not a standards body *per se*, but has acquired significance in recent years in relation to CEPT's activities. CEPT does not admit participation by manufacturing organizations, but depends on EUCATEL to provide the coordinated view of European manufacturers on specific matters under study that are influenced by manufacturing considerations.

10.12 EUROPEAN COUNCIL OF TELECOMMUNICATIONS USERS ASSOCIATION (ECTUA)

European Council of Telecommunications
Users Association
126, avenue Nouvelle
B-1040 Brussels
Belgium

Telephone: 32 2 211 90 06
Facsimile: 32 2 647 23 54

In view of a "Europe of Telecommunications," the European Council of Telecommunications Users Associations was set up in Brussels on 4 March 1986 by the national users' associations of the European Community.

ECTUA represents the common interests of European users to the appropriate organizations; in particular the European Parliament, the CEC, the EFTA, the CEPT, and the ETSI.

Today, ECTUA has ten member associations representing telecommunication users in Belgium, the Federal Republic of Germany, France, Italy, the Netherlands, Portugal, Spain, and the United Kingdom. ECTUA has correspondents in Denmark, Finland, Greece, Ireland, Sweden, Switzerland, and Yugoslavia, and is seeking contacts in other European countries.

Since its creation, ECTUA has produced several position papers about key issues to be contained in a European telecommunication strategy. ECTUA is well recognized as representative of European telecommunication users. In particular, ECTUA organizes several round tables of experts on such subjects as ONP and standards. All the members can participate in User Expert Round Table meetings and in drafting the agreed position. After discussion in such a meeting, the draft position circulates among the members for comments and approval.

As of this writing, 27 individual companies are registered as associate members of ECTUA and this number is continually growing. ECTUA is open to any company that is a user of telecommunication and operating in Europe, which can participate in ECTUA's work as an associate member. Full membership is reserved for national users' associations.

10.13 EUROPEAN TELECOMMUNICATIONS AND PROFESSIONAL ELECTRONICS INDUSTRIES (ECTEL)

Secretariat
EEA
Leicester House
8 Leicester Street
London WC2H 7BN
England

| Telephone: | 44 1 437 0678 |
| Facsimile: | 44 1 437 6047 |

Technical Office
FABRIMETAL
21, rue des Drapiers
B-1050 Brussels
Belgium

| Telephone: | 32 2 510 2434 |
| Facsimile: | 32 2 512 7059 |

The European Telecommunications and Professional Electronics Industries Association (ECTEL) was formally established on 1 November 1985 by the members of the European Conference of Radio and Electronic Equipment Association (ECREEA) and the European Conference of Associations of Telecommunications Industries (EUCATEL). ECTEL's purpose is to act as a joint body on behalf of both conferences in all matters of common concern.

ECTEL includes in its membership the associations from Belgium, Denmark, France, the Federal Republic of Germany, Italy, the Netherlands, Spain, and the United Kingdom, which collectively embrace over 90 percent of the telecommunication and professional electronics companies in the European Community.

ECTEL aims to represent the European industry view on matters of common concern to its members through discussion and cooperation with relevant authorities such as the European Commission and CEPT, both within the European Community and throughout the world.

Organization and Structure

A presidents' conference is held annually for the purpose of electing the president of ECTEL and to approve policy. Other meetings are held as required. Policies are implemented through ECTEL directors' meetings, which normally occur quarterly under the direction of the ECTEL secretary-general.

The president of ECTEL serves until the next presidents' conference, and is supported by a past president and a vice-president, who is the president-elect. The office is held in rotation by countries represented through ECTEL member associations.

The ECTEL secretary-general is appointed for a two-year term by the presidents' conference and may be reappointed; the secretary-general is supported by a secretariat, whose officers are an assistant secretary-general and a technical officer.

ECTEL has a technical organization reporting to the directors comprising four study groups drawn from member companies of ECTEL associations: ECTEL Study Group Telecommunications (ESGT), Mobile Radio Study Group (MRSG), Public Contracts Study Group (PCSG), Regulatory Affairs Study Group (RASG). Each study group comprises a number of subgroups.

Membership

The eight founding associations of ECTEL are full members of the conference. The constitution (protocol) makes provision for telecommunication and electronic equipment associations in EFTA countries to join initially as associate members and to become full members in due time. Provisions exist to permit appropriate European associations from outside the EEC to join ECTEL as observers. At present, associate members are ELIF (Sweden), OeVdEI (Austria), and FFEEI (Finland), and VSMI (Switzerland) is an observer.

ECTEL recognizes changing patterns of the telecommunication and professional industries both within the EEC and is committed to a constant review of its membership criteria and representation.

ECTEL's Activities

ECTEL's activities comprise representation of its members' industrial, technical, and commercial interests in

- public and private telecommunication switching equipment for voice and data services;
- telecommunication transmission equipment;
- public broadcasting equipment and closed-circuit television;
- public and private telecommunication terminal equipment for voice and data services; and
- radiocommunication equipment, radio, and radar navigational aids.

ECTEL acts as a catalyst and forum for discussion of its members' business environment. An example was the formulation of an acceptable *modus operandi* and agreements by 30 companies, members of the ECTEL associations, for European industry participation in the Definition and Main Phases of the European Community on Advanced Communications in Europe. On technical matters, ECTEL's primary influence is in the field of telecommunication and information technology standards, and the harmonization of equipment requirements for new services.

10.14 THE ELECTRONIC AND BUSINESS EQUIPMENT ASSOCIATION (EEA)

> The Electronic and Business Equipment Association
> Leicester House
> 8 Leicester Street
> London WC2H 7BN
> England
>
> Telephone: 44 1 437 0678
> Facsimile: 44 1 434 3477

The Electronic and Business Equipment Association (EEA) merged in 1989 with the Business Equipment Information Technology Association (BEITA) to form the Electronic and Business Equipment Association, but is still known by the initials EEA. The EEA represents 160 manufacturers, including the major British electronic manufacturers, giving a corporate industry view on important technical and commercial matters.

The EEA has a strong mission in information technology and telecommunication standards, and its basic organizational structure in the standards area mirrors the international committees, JTC 1 and CCITT, while in European affairs, EEA is heavily involved in ETSI and CEN-CENELEC. It naturally plays a large part in the formation of national standards by its very heavy role in BSI.

Organization

The EEA has a council to which its director reports monthly, responsible for the broad direction of EEA membership's interest. To support the council and director are eight divisions, each consisting of a permanent staff executive. These divisions are

- Commercial,
- Communications,
- Information Technology,
- Aviation,
- Engineering Services,
- Office Products and Furniture,
- Publicity and Exhibitions,
- European Affairs.

The information technology division is divided into an information technology group dealing with policy and the use of information technology and an information technology standards section, which has 18 committees that deal in such areas as OSI, EFTPOS, telecommunication, EMC, and information technology safety. The standards structure reflects that of BSI, TCT, CCITT, CEN-CENELEC, ETSI, and JTC 1.

10.15 THE EUROPEAN FREE TRADE ASSOCIATION (EFTA)

European Free Trade Association
9-11, rue de Varembé
CH-1211 Genève 20
Switzerland

Telephone: 41 22 734 90 00
Facsimile: 41 22 733 92 91

The aim of the European Free Trade Association is to promote standardization at the European level with a view to facilitating the exchange of goods and services by eliminating obstacles caused by technical requirements. To this end, EFTA develops technical, scientific, and economic procedures necessary to give effect to standardization activities and to cooperate with any international, private, or public organization representative of European and worldwide interest.

The 1960 EFTA convention contained no internal harmonization objective comparable to Article 100 of the Treaty of Rome. Nonetheless, EFTA from the start has considered the elimination of nontariff barriers to internal trade as a prime objective.

EFTA countries have always maintained very strong relationships with CEN and CENELEC, and develop these ties by participating in the procedure for information exchange on standards instituted by EEC Directive 83/189. EFTA provides the financial contribution to this procedure, and supports the ICON database.

Cooperation with the European Free Trade Association

In recognizing the trend toward stronger cooperation, the CEC has proposed in COM(86)547 to open participation in the RACE main program to public or private organizations established in COST (European Cooperation in the Field of Scientific and Technological Research) countries when a framework agreement on research and development cooperation has been signed with the corresponding country. COST comprises all EC and EFTA member countries plus Yugoslavia and Turkey. The general objective of the EEC in telecommunication with regard to its EFTA neighbors is the creation of a wider European economic sphere, and the maintenance and development of the coherence developed in the CEPT framework.

Specific progress has been made in EC-EFTA cooperation concerning

- cooperation on standards and mutual recognition of type approval, and
- cooperation on research and development (RACE program).

10.16 EUROPEAN STRATEGIC PROGRAM FOR RESEARCH AND DEVELOPMENT IN INFORMATION TECHNOLOGY (ESPRIT)

European Strategic Program for Research
and Development in Information Technology
Commission of the European Communities
200, rue de la Loi
B-1049 Brussels
Belgium

Telephone: 32 2 235 20 38
Facsimile: 32 2 235 64 61

The ESPRIT program, commenced in 1984, is a collaborative research and development project bringing together industrial and academic partners across the European community. As a result of annual calls for proposals, selected transnational European consortia work on precompetitive information technology research and development projects, which are co-funded by the EEC.

ESPRIT is a ten-year program (1984-1993) of collaborative research organized in close liaison with industry, national governments, and the research community. ESPRIT is designed to help provide the European information technology industry with the key components of the technology needed to be competitive on world markets within a decade. ESPRIT is an international rather than national program.

Through a community approach, European companies and laboratories can mount a combined research and development drive across national frontiers. This action optimizes the effective management of concurrent research. By bringing together researchers across frontiers, ESPRIT promotes the major transfer of technology throughout the community and opens collaboration between academic and industrial attitudes, scientists, and engineers.

ESPRIT's objectives include

- to provide the European information technology industry with the basic techniques and equipment to meet the competitive requirements of the 1990s;
- to promote European industrial cooperation in information technology; and
- to facilitate the development of standards.

ESPRIT supports precompetitive research and development in

- advanced microelectronics,
- software development,
- office systems,
- computer-integrated manufacturing, and
- advanced information processing.

In ESPRIT, participating companies share the costs and the results of research before competing again at the product development stage. Participants must already be established in the EEC and carrying out research and development information technology. ESPRIT participants also must have a partner in another EEC country.

While the ten-year program sets technical guidelines, the research follows a more detailed workplan, which is drafted each year according to the monitoring of progress in information technology, both within the community and worldwide. The workplan is prepared by some 250 experts in the field, who are not a fixed group, in consultation with industry and governments so that the program is in constant touch with reality. Each year current work on ESPRIT projects is evaluated. There is an ESPRIT Management Committee (EMC) and an ESPRIT Advisory Board (EAB). The draft workplan, after approval by EMC and EAB, is adopted by the CEC and sent to the EC for approval. On the basis of the workplan, a new call for proposals is then made early the following year in the form of an open invitation published in the official journal of the European Communities.

ESPRIT projects submit contracts along the technical lines detailed in the workplan. The projects should demonstrate:

- technical soundness;
- contribution to industrial strategy with respect to ESPRIT objectives;
- community dimension;
- technical, scientific, and managerial capacity to carry out the proposed program of work; and
- measures envisaged and approaches to accessibility and exploitation of results.

Proposals are either for type A projects, which are "system-driven," which direct large teams of specific technology goals and form the backbone of the ESPRIT program. Type B projects are "idea-driven," relying more on flexible infrastructure and latitude in individual thinking. The CEC, after consultation with the EMC and EAB, awards the contracts and supervises their execution.

Daily management is through the CEC's Task Force for Information Technology and Telecommunications, which reports directly to the member of the commission responsible for industrial affairs.

Researchers are linked by the ESPRIT information exchange system, an advanced electronic network. Workshops on each research topic provide a detailed review of problems and progress.

In 1990, some 450 new projects for a total cost of 3.9 billion ECU have been submitted to the ESPRIT program. Half of the projects concern information processing systems (IPS), 28% computer integrated manufacturing (CIM), and 22% office and business systems (OBS). The CEC, taking into account the advice of the ESPRIT Review Board, is likely to grant increased participation by small and medium sized companies. In this second phase, some 6,000 researchers will be working in a strong network of scientists on key topics relating to ESPRIT goals.

A comprehensive independent review of the ESPRIT program found that, in the vast majority of projects, trans-European cooperation has been a success, resulting in significant benefits for both the participants and Europe's technological base. Techniques, facilities, and human resources have all improved, and good work has been done on international standards. Links between industry and universities, managerial awareness of the strategic importance of information technology, and optimism about the future have been increased.

10.17 THE RACE PROGRAM

RACE Industrial Consortium
61, rue de Trèves
B-1040 Brussels
Belgium

Telephone:	32 2 236 35 04
Facsimile:	32 2 236 29 81

RACE is the acronym for R&D in Advanced Communication Technologies for Europe.

In July 1985, the EC authorized an 18-month RACE definition phase. In October 1986, with the definition phase drawing to a close, the CEC submitted proposals and a workplan for a RACE main phase. RACE aims at the introduction of integrated broadband communication (IBC), taking into account the evolving ISDN and national introduction strategies, progressing to Community-wide services by 1995. The prenormative (or prestandardization) activities avoid the familiar difficulties of *ex post factum* harmonization and ensure the interoperability of future systems.

An IBC reference model, the basic design of European broadband network and services, was elaborated by three groups:

- "networks," which defined standards requirements, and was handled by GSLB (*Groupe Spécial Large Bande*) of CEPT;
- "terminals," involving shared-cost contracts with manufacturers, research laboratories, and telecommunication administrations; and
- "services," through the SOG-T/GAP.

The CEC has also studied the transnational broadband backbone (TBB), which is to accelerate the establishment of high-bandwidth cross-frontier transmission links.

RACE Proposal 1087 is designed to make provision of verification (PROVE). The PROVE proposal was submitted to the RACE central office as a follow-up to RSVP, a previous project titled RACE Strategy for Verification and Plan. The latter was executed by a consortium of PTTs representing industry and users with SPAG as prime contractor during 1988. PROVE negotiations have been completed, and work currently underway will span a four-year period (1989–1992).

The PROVE consortium is composed of SPAG (prime contractor), British Telecom, Deutsche Bundespost, France Télécom, EOLAS Irish Science and Technology Agency, Televerket, Standard Elektrik Lorenz/AL-CATEL, CAP REGION, Clemessy, Elektronik Centralen, Hasler and Refer. NCC is participating as consultant to the project. The CEC will contribute up to 50 percent of the 13.6 million ECU budget. SPAG contributes with 95 person-months over four years.

The objectives of the proposal are

- taking a leading role to establish consensus on broadband ISDN (B-ISDN) verification concepts, requirements, and procedures;

- establishing methods and procedures for interoperability testing in the B-ISDN environment;
- defining a methodology and recommendation for computer-aided test case generation; and
- studying the feasibility of the development of prototype verification tools for

—system access equipment,

—traffic load generator, and

—signaling control units.

10.18 COOPERATION FOR OPEN SYSTEM INTERCONNECTION NETWORKING IN EUROPE (COSINE)

COSINE is a program for a pan-European research computer network. COSINE aims to build a single network across the continent, linking numerous university and laboratory networks, currently in a pilot phase.

COSINE is seen as crucial to the future of European research. Currently, links among various research groups are limited and disjointed. COSINE aims to overcome this problem by providing a 64-kilobyte-per-second X.25 carrier network supporting services such as X.400 message handling, remote (virtual terminal) access, and OSI file transfer access and management (FTAM). COSINE has a number of gateway development projects for meeting the needs of established networks based on Transmission Control Protocol–Internet Protocol (TCP/IP), DECnet, and IBM Systems Network Architecture (SNA). Each gateway will be connected to a particular non-OSI network type. COSINE would then support the non-OSI protocols without allowing them to run on the network.

Working group NA6 has recently started work on standards for intelligent networks (INs).

10.19 THE STAR PROGRAM (STAR)

The Special Telecommunications Action for Regional Development (STAR) Program

The STAR is a five-year program (1987–1991) aimed at using advanced telecommunication services to promote the economic development of the less favored regions of the European Community. The STAR program is active in France, Greece, Ireland, Italy, Spain, Portugal, and the United

Kingdom, although some of these countries have only designated regions eligible for assistance.

The program is designed to

- help set up the telecommunication infrastructure necessary to provide advanced services to business users in the less favored regions; and
- stimulate demand and encourage use of telecommunication infrastructure in less favored regions.

Because the objective is economic growth, STAR is specifically conceived for business users with special emphasis on small and medium sized enterprizes. The European Community recognizes the key role of smaller enterprizes in the economic fabric of Europe and that such enterprizes have particular needs, which must be identified and catered.

Recently, the commissioner for regional policies approved a number of regional development programs, projects, and other measures, focusing on the adoption of the *Community Support Frameworks* for regions where development is lagging behind (Objective 1) and regions in industrial decline (Objective 2). The operational program shows Objective 1 regions in Spain, Ireland, Italy, Portugal, and the United Kingdom (Northern Ireland), and those for Objective 2 in the United Kingdom, France, and the Federal Republic of Germany. These programs are financed by the European Community funds.

Mobile Communication

In February 1987, the CEC submitted two proposals to the EC:

- for the coordinated introduction of public pan-European digital mobile communication in the EEC, and
- on the frequency bands to be made available for this purpose.

Member states are to reserve frequencies for a second-generation pan-European digital system, which is the major precondition for a future Community-wide system.

10.20 TRADE ELECTRONIC DATA INTERCHANGE SYSTEMS (TEDIS)

In December 1986, the CEC proposed a program on Trade Electronic Data Interchange Systems (TEDIS). International trade requires a vast volume of data exchange between business partners, and the data exchange may be more time-consuming than the manufacture or delivery of the traded goods or services. Standardized Electronic Data Interchange (EDI) has developed as one of the most promising lines of value-added services, based on cooperative ventures, mainly established by industry associations.

10.21 AREAS OF TELECOMMUNICATION STANDARDIZATION

Integrated Services Digital Network (ISDN)

In December 1986, the EC issued a recommendation for coordinated introduction of ISDN. The recommendation set an objective for 1993 market penetration for a number equivalent to 5 percent of main lines, and an objective for 1993 availability, with 80 percent of customers having the option of ISDN access. This recommendation does not require substantial research and development because it is based on the ongoing digitization of the telephone network. ISDN is to be the EEC's open network infrastructure.

The CEC is attempting to provide access to ISDN for subscribers in most parts of Europe in the early 1990s. Services offered by ISDN represent

- increased revenue from new services and growing use of the network for operating companies,
- increased market volume for new equipment for industry, and
- new and improved telecommunication systems for users.

Full acceptance of the ISDN service by telecommunication users will be achieved only if the costs are comparable to those of services over existing dedicated networks. The key to reducing the cost of the new service is international standardization. To ensure ISDN's standardization the EC issued a recommendation on the targets for introduction of ISDN (Phase 1) in Europe from 1983, which should involve

- a natural evolution of the telephone network, giving ISDN subscriber access at 144 kilobits per second and 2 megabits per second;
- a standard physical interface between ISDN terminals and the public network;

- a universal telecommunication plug and socket terminal connection;
- priority on a number of ISDN services including:
  —numbering, addressing, and signaling,
  —tariff considerations,
  —interworking between national ISDN trials,
  —a high level of penetration, and
  —integration with digital cellular mobile networks and broadband services.

Open Network Provision (ONP)

At present, the provision of pan-European services is often made technically or administratively impossible by the absence of harmonized technical interfaces, divergent conditions of use, or discriminatory tariff principles. The ONP directive aims to eliminate these anomalies through harmonization in close collaboration with the European Telecommunications Standards Institute (see Table 10.8). Under the ONP directive:

- technical interfaces and service features will become the subject of European standards adopted by ETSI. Service providers complying with such standards will be able to offer their services throughout the European Community; and
- the CEC is empowered to make standards mandatory to the extent necessary to ensure interoperability of transfrontier services within the Community and improve freedom of choice for users. Mandatory standards are unlikely for value-added services because the procedure outlined above was designed for application to basic services such as packet-switched data transmission and ISDN.

The ONP directive provides a framework for directives on specific issues, such as

- leased lines and voice telephony,
- technical interfaces and service features for packet-switched data transmission and ISDN,
- recommendations for packet-switched data transmission and ISDN, and
- CEC proposals to be made into directives.

Table 10.8. ETSI Structure and the Role of ISDN Systems Management (ISM) (courtesy of ETSI ©1989 Editions Rouland)

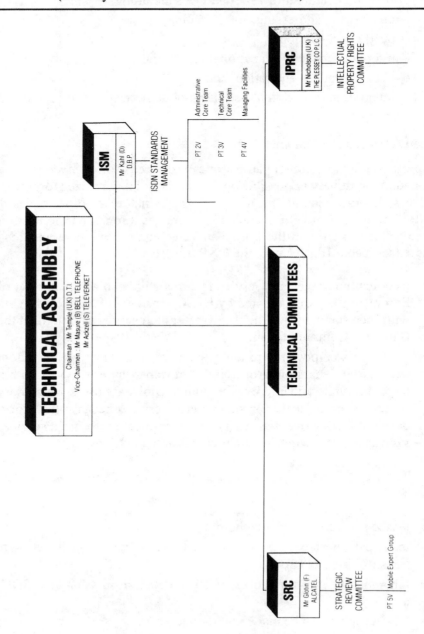

The services directive is a parallel initiative, aimed at liberalization of telecommunication services, under which

- the exclusive or special rights of PTTs in telecommunication services are to be abolished with the exception of voice telephony and network infrastructure. At the end of 1992, all member states must permit simple resale of capacity of leased lines.

Thus, private service providers will offer value-added telecommunication services in competition with the PTTs and, from the beginning of 1993, resale of basic services.

Pan-European Mobile Communication

CEPT has made a vigorous start on using the transition to second-generation technology, as a chance to develop a common European system of digital cellular mobile communication. A special working group, the Special Mobile Group (GSM), was set up by CEPT for this purpose. CEPT members agreed to reserve the frequencies 905–915 MHz and 950–960 MHz for the introduction of future pan-European mobile communication services. CEPT members must employ these reserved frequencies before yielding to pressure by national groups to increase the bandwidths used by current national or regional systems.

The Senior Officials Group for Telecommunications (SOG-T) set up a Group for Analysis and Forecasting (GAP), which looked into the question of pan-European mobile communication. In June 1987, the EC issued a recommendation and a directive on pan-European mobile communication.

Because of the fortunate timing of the directive, the CEC will give high priority to terminal specification, particularly, the establishment and mutual recognition of type approval for telecommunication terminal equipment and the appropriate NETs.

10.22 ASSOCIATION FRANÇAISE DE NORMALISATION (AFNOR)

Association Française de Normalisation
Tour Europe
Cedex 7
F-92080 Paris la Défense
France

Telephone: 33 1 42 91 55 55
Facsimile: 33 1 42 91 56 56

A "permanent standardization commission" was created by decree in 1918. This organization did not survive, however, and the Association Française de Normalisation (AFNOR) was founded in 1926 to fill the resulting void. AFNOR is a private, nonprofit association under the French law of 1 July 1901, and is recognised as a public service by the decree of 5 March 1943. By the decree of 26 January 1984, the Ministry of Industry and Research made AFNOR responsible for standardization in France.

AFNOR has a staff of 144 directly employed by the member body and a staff of 250 working for the member body but sponsored by other organizations. AFNOR's primary responsibilities include the preparation of standards, sale of publications, marking of goods, quality control services, metrology, education, promotion, and testing facilities. AFNOR has no testing responsibility, but, for purposes of developing standards and for certification, the association is assisted by a great many specialized, officially recognized laboratories.

10.23 BRITISH STANDARDS INSTITUTE (BSI)

British Standards Institute
2 Park Street
London W1A 2BS
England

Telephone: 44 1 629 90 00
Facsimile: 44 1 629 05 06

The British Standards Institute (BSI) developed from the Engineering Standards Committee, formed in 1901. The present title was adopted in 1931. A royal charter was granted in 1929, under which the institute

- coordinates the efforts of producers and users for the improvement, standardization, and simplification of engineering and industrial materials so as to simplify production and distribution and to eliminate the "national waste" of time and material involved in the production of an unnecessary variety of patterns and sizes of articles for one and the same purpose;
- sets up standards of quality and dimension, and prepares and promotes the general adoption of British standards specifications and

schedules in connection therewith and revises, alters, and amends such specifications and schedules as experience and circumstances may require;

- registers, in the name of the institute, marks of all descriptions, and proves the affixing or licenses the affixing of such marks or other proof, letter, name, and description; and
- takes such action as may appear desirable or necessary to protect the objects or interests of the institute.

The UK Secretary of State for Trade and the president of BSI (on 24 November 1982) signed a Memorandum of Understanding on standards, covering the preparation, application, and promotion of British standards. The electrotechnical council of BSI forms the British Electrotechnical Committee, the national committee of the IEC for the United Kingdom.

BSI has a staff of 1,236 directly employed by the member body and in addition three staff members work for the member body but are sponsored by another organization.

10.24 DIN DEUTSCHES INSTITUT FÜR NORMUNG

Deutsches Institut für Normung
Burggrafenstrasse 6
Postfach 1107
D-1000 Berlin 30
Germany

| Telephone: | 49 30 260 01-1 |
| Facsimile: | 49 30 260 12 31 |

DIN Deutsches Institut für Normung is a private, nonprofit association founded in 1917. Relations between DIN and the government, both on the federal and the *Länder* (state) levels, are regulated by contracts.

DIN standards are recognized by industry, trade, labor unions, consumers, government, and lawyers as accepted rules of technology. Principles which DIN follows are that

- standards are voluntary in nature,
- standards projects and drafts are made available for public comment,
- all interested parties can participate,
- DIN standards form a unified and consistent whole,

- the standards keep to the technical matter at hand,
- the standards are geared to technological development,
- the standards are matched to economic conditions,
- the standards should benefit the community as a whole, and
- world trade should be free of technical barriers.

To promote the implementation of standards, DIN organizes training courses, regular exchange of experience among standards practitioners, and also participates in certification and quality assurance. DIN's publishing house, Buth-Verlag, provides 60 percent of DIN's total budget through its sales of standards and associated technical literature in printed form, on microform, and on electronic media.

DIN has a staff of 590 directly employed by the member body, and an additional staff of 180 working for the member body but sponsored by other organizations.

CHAPTER 11

ASIAN AND PACIFIC ORGANIZATIONS

11.1 INTRODUCTION

This chapter considers organizations which are representative of a number of countries in the Asian and Pacific region. The chapter then considers Japan, first with an introduction on regulation in that country and next with a study of organizations involved in telecommunication standards. Also, certain organizations from Japan involved in telecommunication standards are discussed in Chapter 12, where functional standardization and some applications development of telecommunication standards are covered.

This chapter also considers telecommunication standards organizations in Australia and New Zealand, and finally itemizes Asian countries which have member organizations of ISO, and describes those organizations. In Australia, the quintet now appearing in some countries is illustrated. These are a CCITT committee, a JTC 1 committee, a users' group, an industry group, and an independent regulator committee.

11.2 PACIFIC TELECOMMUNICATIONS COUNCIL (PTC)

Pacific Telecommunications Council
1110 University Avenue, Suite 308
Honolulu, Hawaii 96826

Telephone: 808 941 3789
Facsimile: 808 944 4874

The Pacific Telecommunications Council (PTC) was set up in 1980 as a voluntary, independent organization to meet the growing need for "the development, understanding, and beneficial use of telecommunications in the Pacific area."

The PTC, located in Honolulu, where annual conventions are held, has a large membership, which includes most of the leading companies in

---

telecommunication in the United States and Japan.

The stated objectives of the PTC are to

- provide a forum for discussion and interchange of information, views, and ideas on telecommunication in the Pacific area by bringing together a multifaceted and diverse body, including users, planners, and providers of telecommunication services and equipment;
- promote a general awareness of telecommunication needs in the Pacific area;
- organize conferences, workshops, and seminars to promote interchange and address specific issues; and
- communicate the views and recommendations of the PTC to the established national, regional, and international bodies responsible for telecommunication policy.

11.3 ASIA PACIFIC TELECOMMUNITY (APT)

Asia Pacific Telecommunity
12/49 Soi 5 Chaengwattana Road
Thungsonghong Bangkhen
Bangkok 10210
Thailand

Telephone: 662 573 0044
Facsimile: 662 573 7479

The Asia Pacific Telecommunity (APT) was established through an intergovernmental agreement under the auspices of the United Nations Economic and Social Commission for Asia and the Pacific (ESCAP). APT was formed to ensure balanced development of telecommunication in the Asian and Pacific region at a pace commensurate with the economic and social development of the area. Membership is open to any state within the region that is a member of the United Nations or ESCAP. Member countries include Afghanistan, Australia, Bangladesh, Burma, Brunei, Darussalam, the People's Republic of China, India, Indonesia, Iran, Japan, Maldives, Malaysia, Nauru, Nepal, Pakistan, the Philippines, Republic of Korea, Singapore, Sri Lanka, Thailand, and Vietnam.

APT provides such functions as

- the programming, implementation, and coordination of technical standards on a regional basis for the Asian telecommunication infrastructure;

- coordination of commercial relations between telecommunication administrations of the region;
- provision of advice on economic and rational operating methods in national and regional telecommunication services;
- assistance in evaluation of the needs of telecommunication administrations for training of personnel and in the transfer of telecommunication technology with the aim of improving the efficiency of the telecommunication services in the Asian and Pacific region; and
- provision of short-term technical assistance to support and supplement the assistance provided by the ITU, the United Nations Development and Assistance Program (UNDAP), and other bilateral or multilateral organizations.

APT undertakes technical assistance and other studies relating to telecommunication networks and related technological developments, regularly organizes seminars and workshops to collate planning, programming, and reviewing of technological developments, and provides a forum for the exchange of information. APT also serves as a consultative organization for telecommunication matters which are more effectively discussed and resolved on a regional basis.

11.4 JAPANESE MINISTRY OF POSTS AND TELECOMMUNICATIONS (MPT)

General Planning and Policy Division
Minister's Secretariat
3-2, Kasumigaseki 1-chome
Chiyoda-ku, Tokyo 109
Japan

Telephone:	813 504 4947
Facsimile:	813 509 4565

Japan's Ministry of Posts and Telecommunications (MPT) is responsible for the supervision of the country's postal, postal banking, post office life insurance, telecommunication, and broadcasting services.

New Carriers Offering Telecommunication Services

In April 1985, the government reformed the telecommunication sector in Japan, introducing competition and privatizing the Nippon Telegraph and

Telephone Corporation (NTT). Since then a variety of telecommunication carriers have entered the market. As of 1989, there were 46 so-called Type I telecommunication carriers, 26 special Type II telecommunication carriers, and 688 general Type II telecommunication carriers.

A number of new carriers entered the Type I field of long-distance, regional, satellite, automobile telephone, radiopaging, and international telecommunication. In the Type II field, many carriers were newly established in various parts of Japan to help financial, manufacturing, distribution, and transportation corporations to set up their own telecommunication networks.

ISDN

An ISDN was inaugurated in Tokyo, Osaka, and Nagoya on 19 April 1988. At the end of December 1988, 877 leased circuits were serving 181 companies in the fields of finance, manufacturing, and information services, covering 28 geographic areas including Fukuoka, Sapporo, and other designated cities.

New Type I Telecommunication Carriers

Companies in Japan's Type I category of telecommunication carrier install and operate their own telecommunication circuit facilities, providing service in the long-distance, satellite, regional, automobile telephone, radiopaging fields, and more. By the end of fiscal 1988, forty-three companies had entered this category, forty-one of which operated domestic communication services and two of them operating internationally.

Of the thirty-four carriers now operating, five offer both telephone and leased circuit services, three offer leased circuit services exclusively, two offer automobile telephone service, and twenty-four offer radiopaging services.

By the end of 1988, eight carriers provided leased circuit services. At the end of 1988, twenty-four companies were offering radiopaging. Only two long-distance telephone services and three radiopaging services were profitable by the end of fiscal 1988. The aggregated revenues of the three long-distance telephone services for the first half of fiscal 1988 were equal to only 1.1 percent of NTT's telephone revenues, showing that true competition had not yet been achieved.

New Type II Communication Carriers

During fiscal 1988, 163 new Type II carriers entered the telecommunication field by leasing circuits from other carriers. These new carriers offer a wide range of services from leased circuits to value-added networks (VANs). Their operation has contributed to the installation of networks in the financial, manufacturing, distribution, and transportation sectors and realization of the advanced information-oriented society.

There are two categories of Type II carriers. Special Type II telecommunication carriers, offering international or large-scale domestic services, numbered 25 at the end of fiscal 1988. On 1 July 1988, NTT separated its data communications sector, and the NTT Data Communications Systems Corporation was established as its successor, becoming one of the largest special Type II carriers in Japan. If this company's revenues are aggregated, NTT's total revenue rose 3 percent from the previous year.

General Type II carriers, numbering 668 at end of fiscal 1988, offer mainly transmission and switching services involving format, code, or protocol conversion, for such purposes as sending and receiving orders; other services include personal computer networks and voice-message storage services.

New International Type I and Type II Telecommunication Services

Kokusai Denshin Denwa Co. Ltd. (KDD), Japan's major international telecommunication carrier, was the main recipient of the increase in international telephone circuits, which reached 255 million in fiscal 1988, up 34 percent from the previous year. The company's sales were up 5 percent and ordinary income up by nearly 11 percent.

In November 1987, two companies, International Telecom Japan, Inc. (ITJ), and International Digital Communications, Inc. (IDC), received permission from MPT to offer Type I international leased circuit services. Both companies began international telephone services late in 1989, offering services to ten countries, centering on advanced Western nations.

The offering of international VAN services was realized by amendments to the Telecommunications Business Law in September 1987, and 13 companies are now registered as international special Type II telecommunications carriers. At the end of fiscal 1988, 12 companies had made contracts with foreign partners and entered the international telecommunication business field, particularly with services between Japan and the United States and between Japan and the United Kingdom.

Technological Development in Telecommunication

On 26 October 1988, the MPT approved establishment of the Support Center for Advanced Telecommunication Technology Research to conduct surveys and to support joint international research and development. The MPT promoted ISDN standards in the CCITT and held an OSI-ISDN promotion conference. To confirm capability of systems and terminals on the basis of ISDN standards, there must be a valid means of testing mutual connectivity. The MPT is sponsoring harmonization of advanced telecommunication systems (HATS) conferences to resolve such problems.

The MPT has also pursued work on intelligent communication networks, which will provide a variety of high-level services like "fuzzy access," which enables connection to a desired receiver through fuzzy (i.e., plain language) information such as key words, rather than conventional telephone numbers. Other applications include personal telephone number services, data storage, and switching functions with natural language processing. Work has also proceeded on artificial intelligence (AI) in telecommunication.

Three subcommittees of the TTC (see Section 11.5) are looking at the future of the telecommunication industry. The first subcommittee is deliberating on a suitable structure for the telecommunication market; the second is studying the form that services and the rate system should take on ISDN; the third is considering the appropriate role of NTT. A round table conference on communication policy has also been set up as an advisory body to MPT.

In satellite communication, MPT is promoting the Broadcasting and Communications Technology Satellite (BCTS), which is being researched by MPT's Space Development Committee. This committee is also looking at the Experimental Data Relay Tracking Satellite (EDRTS) technology.

11.5 THE TELECOMMUNICATIONS TECHNOLOGY COUNCIL (TTC)

> Telecommunications Technology Council
> Hamamatsuchou-suzuki Building
> 1-2-11, Hamamatsu-chou
> Minato-ku, Tokyo 105
> Japan

Telephone: 81 3 432 1551
Facsimile: 81 3 432 1553

The Telecommunications Technology Council (TTC) was established in April 1985 as a result of the merger of the Radio Technology Council and the Sectional Meeting for Technology of the Telecommunications Council, both organs of the MPT. The TTC was founded to provide a forum for deliberations on the rapid development of telecommunication technology, both wireline and wireless, and to determine appropriate means of carrying out telecommunication policies. Ultimately, the TTC acts as an advisory organ to MPT on telecommunication technology and policy.

The TTC attempts to ensure transparency in setting standards and technical specifications, in accordance with the September 1985 "Guidelines for Ensuring Transparency in the Process of Drafting and Revising Standards," enacted by the Japanese government's Action Program Promotion Committee.

The main decision-making body of the TTC is the advisory committee. Its membership consists of 20 individuals from the academic and business worlds chosen by MPT. The advisory committee is divided into 17 committees, one for each of the organizations with which the TTC has relations. Some of the committees are further divided into expert committees, which deliberate matters of technology and standardization; the expert committees have about 200 members. Subcommittees to the committees and expert committees are formed as needed. There are approximately 500 subcommittee members; these are researchers who study technical matters and report on them to their respective committee or expert committee. The advisory committee meets once a month.

Committee members, expert committee members, and technical researchers are from government ministries and agencies, local government, universities, NTT, KDD, Japan Broadcasting Corporation (NHK), telecommunication manufacturers, public service corporations, carriers, distributors, electric power companies, and telecommunication research institutions.

The TTC acts as a clearing house for data on standards and technical specifications from international organizations, reporting such data to equipment manufacturers and telecommunication carriers. Likewise, opinions and advice on these and other matters are transmitted from private sector groups to international organizations by the TTC. International organizations with which the TTC maintains relations include the CCITT and CCIR of the ITU. The International Maritime Organization (IMO), International Civil Aviation Organization (ICAO), and Comité International Spécial des Perturbations Radioélectriques (CISPR) of the International Electro-Chemical Commission (IEC).

The TTC establishes technical requirements for telecommunication and broadcasting. Technologies covered include teletext, aircraft public telephones, electronic mail, information processing systems, cordless telephones, pagers, automatic radio monitoring systems, cellular telephones, optical communication, technological measures for security and reliability of telecommunication systems, geostationary satellites, multichannel cable television, coastal radiotelephones, high definition television, artifical intelligence, and other specific matters.

11.6 JAPANESE INDUSTRIAL STANDARDS COMMITTEE (JISC)

Japanese Industrial Standards Committee
Standards Department
Agency of Industrial Science and Technology
Ministry of International Trade and Industry
1-3, Kasumigaseki 1-chome
Chiyoda-ku, Tokyo 100
Japan

| Telephone: | 81 3 501 92 95/6 |
| Facsimile: | 81 3 380 14 18 |

The Japanese Industrial Standards Committee (JISC) was established by the Industrial Standardization Law in 1949 as an advisory organization to ministers in charge of Japanese industrial standards (JIS) and the marketing of products with the JIS mark.

The JISC consists of the General Meeting, the Standards Council, 29 Divisional Councils, and 1,008 Technical Committees, which have producers, dealers, users, consumers, and academics as their members. About 8,200 standards cover all fields of industrial and mineral products, except medicines, agricultural chemicals, and chemical fertilisers, which are governed by other laws.

The JISC has a staff of 92 directly employed by the Standards Council and 155 staff persons working for the council but sponsored by other organizations. The primary responsibilities of the JISC include the preparation of standards, marking of goods, education, promotion, and industrial research.

11.7 JAPANESE STANDARDS ASSOCIATION (JSA)

Japanese Standards Association
1-24, Akasaka 4-chome
Minato-ku, Tokyo 107
Japan

| Telephone: | 81 3 583 8001 |
| Facsimile: | 81 3 586 2014 |

The Japanese Standards Association was founded as a nonprofit institution on 6 December 1945, under government authorization. JSA's objective is to propagate industrial standardization and quality control to all industry. The six activities which JSA pursues are

- publication and distribution of the (JIS), the English edition of JIS, monthly magazines, related books, standards materials, and other publications;
- propagation of standardization and quality control;
- education and consultation on standardization and quality control;
- study and research of standardization and various engineering techniques; and
- cooperation in international standardization activities.

JSA's income totals 1.1 billion yen, which it derives from the sale of publications and other sources such as lecture fees. There are 8,600 supporting member companies and 4,700 subscriber member companies, supporting members receiving discounts to certain publications and seminars, and subscribing members receiving newly established or revised JIS each month.

JSA has a staff of approximately 160 persons, and a board of directors comprising a president, counsellors, advisers, and executive directors as well as a director-general. The head office has four departments and four centers:

- General Affairs Department;
- Standards Engineering and Inspection Department (divided into a Standardization Division and an Accredited Factory-Inspection Division);
- a Publication Department;
- a Business Department;
- Overseas Standard Center;

- Management Engineering Center (MEC);
- International Standardization Cooperation Center (ISCC); and
- Information Technology Research and Standardization Center (ITRSC).

There are also seven branch offices throughout the nation.

11.8 JAPAN ELECTRONIC INDUSTRY DEVELOPMENT ASSOCIATION (JEIDA)

Japan Electronic Industry Development Association
Kikai-shinko Kaikau
3-5-8, Shiba Koen 3-chome
Minato-Ku, Tokyo 105
Japan

Telephone:	81 3 433 6296
Facsimile:	81 3 433 6350

The Japan Electronic Industry Development Association (JEIDA) was formed in 1958 as a nonprofit organization interested in contributing to Japan's economic prosperity by stimulating development in the electronics industry.

JEIDA members are Japan's leading manufacturers of computers and related equipment. There are nearly 100 regular corporate members and 80 associate members, who work toward common objectives by serving on various boards and committees.

The association's objectives are to develop and improve the various fields in the industry. JEIDA's focus is on such products as computers, automation equipment, information systems, high performance electronic components, and materials.

The JEIDA organization has sixteen directors and two auditors, all of whom are elected at a general meeting. The directors then choose a chair, five vice-chairs, an executive director, and a managing director to serve two-year terms. Nearly forty people work full time in JEIDA'S offices. There are five divisions:

- General Affairs;
- Planning and Coordination;
- Information Industry;

- Fundamental Technology; and
- International Relations.

JEIDA's major responsibilities are to:

- research and analyze the trends or markets and technology both in Japan and elsewhere, compiling market statistics for personal computers, small business computers, minicomputers, terminals, and peripheral equipment as well as to assess long-term economic prospects;
- survey new computer terminals and peripheral equipment, assessing their commercial possibilities;
- provide information on fundamental fields within the electronics industry (e.g., new electronic materials, new devices, and new sensors);
- contribute to the standardization of Japanese information processing, office automation equipment, and electronic medical equipment; and
- conduct surveys aimed at establishing both a Japanese and an international industrial standard.

11.9 JAPAN INFORMATION PROCESSING DEVELOPMENT CENTER (JIPDEC)

Japan Information Processing Development Center
Center for the Informatization of Industry (CII)
3-5-8, Shibakoen
Minato-ku, Tokyo 105
Japan

JIPDEC is the Japanese consultative organization for ISO TC 154 and promotes electronic data interchange (EDI) in Japan.

11.10 STANDARDS ASSOCIATION OF AUSTRALIA

Standards Australia
Standards House
80 Arthur Street
North Sydney, NSW 2060
Australia

Telephone: 61 2 963 4111
Facsimile: 61 2 959 3896

Standards Association of Australia was formed in 1922, but reorganized in 1929 and again in 1950 when it was incorporated by Royal Charter. In 1988 it was recognized as the peak standards-setting body for Australia under Memoranda of Understanding with the Commonwealth of Australia and now operates as *Standards Australia*. Standards Australia's affairs are managed by a council, which comprises representatives of Commonwealth and state governments, government departments, associations of manufacturing and commercial interests, professional institutes, and community and consumer interests. Approximately 20 standards boards oversee the development of standards in designated fields by allocating them to almost 1,700 technical committees.

Standards Australia is a member of ISO and IEC and other relevant regional and international standards-setting bodies.

11.11 AUSTRALIAN CCITT COMMITTEE

Australian CCITT Committee
Telecom Australia
Office of the Corporate Secretary
199 William Street
Melbourne, Victoria 3000
Australia

Telephone:	61 3 606 6350
Facsimile:	61 3 670 2562

Australian CCITT Committee is established as part of Australia's membership in the ITU, and the management of that membership by the responsible government department, the Department of Transport and Communications. The Australian CCITT Committee is composed of representatives from that government department, carriers, AUSTEL, key industry bodies, and Australian standardization organizations. Telecom Australia acts as the chair and the secretariat for the Australian committee.

The committee is responsible for administering Australia's CCITT activities within the framework set by the Department of Transport and Communications, and advises the department on matters concerning the CCITT. The committee acts by direct consideration of relevant issues, and by the establishment and composition of national CCITT study groups to coordinate detailed technical activity. The committee monitors the activities of regional and national telecommunication standards bodies outside Australia, and liaises with the AUSTEL Standards Advisory Committee (AUSTEL-SAC) as appropriate.

National CCITT Study Groups

For each international CCITT study group, a national CCITT study group has been established by the Australian CCITT Committee, which appoints a convenor from a member organization active in the CCITT study group.

Each national CCITT study group holds regular meetings to brief interested organizations on progress, to consider Australian contributions, and to brief Australian delegates (if any) to meetings of the group.

The convenor of a national CCITT study group provides reports to the Australian CCITT Committee on both the main activities of the study and those of other related bodies (such as regional or national standards organizations).

Each national CCITT study group is open to all organizations associated with the Australian CCITT Committee, and to other interested organizations as agreed by the Australian committee.

11.12 AUSTRALIAN STANDARDS COMMITTEE (AUSTEL-SAC)

> Australian Standards Committee
> AUSTEL
> 5 Queens Road
> Melbourne, Victoria 3004
> Australia
>
> Telephone: 61 3 828 7300
> Facsimile: 61 3 820 3021

To allow transition from Telecom Australia's (the government-owned internal carrier) specifications and regulations to AUSTEL, as directed by the Telecommunications Act 1989, Australian Telecommunications Authority (AUSTEL), the recently formed independent regulator determined that Telecom Australia's regulatory specifications 1 to 16 should form the basis of technical standards for Australian telecommunication. This decision was made by AUSTEL after consultation with industry, as an interim measure, with a view to the standards being reviewed by its Standards Advisory Committee (SAC). Nine subsidiary working groups have been formed, comprising members of interested organizations and industry groups. Each working group reviews a standard referred to it by the SAC and reports to the SAC:

- whether a standard is necessary;
- if so, whether it may be improved and how; and
- whether there should be a voluntary component in addition to mandatory requirements.

The subsequently revised standard, recommended by the working group to the SAC, has the objectives of

- protecting the integrity of telecommunication networks and the safety of people working on or using the services;
- ensuring interoperability of customer equipment or customer cabling to telecommunication networks;
- complying with recognized international standards concerning the interface of customer equipment or customer cabling to telecommunication networks; and
- being expressed in "plain English" and in terms consistent with the Telecommunications Act of 1989.

A member of a working group may submit a minority report outlining in what respects and why the member does not accept the majority report. The nine working groups are

- electrical safety and "earthing" (grounding);
- public telecommunication network interface;
- non-ISDN switching;
- cellular mobile telephone;
- telex;
- cabling, wiring rules, and cable licensing requirements;
- ISDN;
- video; and
- 2-megabit-per-second leased lines.

11.13 AUSTRALIAN INFORMATION INDUSTRY ASSOCIATION (AIIA)

Australian Information Industry Association
12 Campion Street
Deakin, ACT 2600
Australia

Telephone:	61 62 82 4700
Facsimile:	61 62 85 1408

AIIA represents all sectors of Australia's information industry. Member companies include manufacturers and suppliers of hardware, and producers of software and related services for computer and telecommunication systems.

AIIA is dedicated to ensuring that the best possible environment exists for the development, production, and adoption of information technologies in Australia. The association provides members with a forum for discussion of issues, and a vehicle for communicating positions on these issues to government and the public.

AIIA members have combined gross annual sales revenues in Australia of almost $A7.4 billion, and employ over 30,000 Australians.

11.14 AUSTRALIAN TELECOMMUNICATION USERS GROUP (ATUG)

Australian Telecommunication Users Group
Sentry Building, Seventh Floor
61 Lavender Street
Milson's Point, NSW 2061
Australia

Telephone:	61 2 957 1333
Facsimile:	61 2 925 0880

ATUG is a nonprofit organization designed to promote the interests of all telecommunication users in Australia to facilitate the availability of a wide range of reliable services and products of quality at affordable prices.

ATUG's main objectives include the development of

- telecommunication services, policies, equipment, standards, and regulations; and
- a viable, technologically advanced and self-sufficient Australian telecommunication industry.

ATUG has an executive based in Sydney and about 500 members, ranging from individuals to large corporations and government departments. ATUG provides a range of services and seeks to assist its user members by

- providing a mechanism for the identification and definition of telecommunication problems;

- liaising with government and telecommunication organizations;
- determining efficient and cost-effective solutions for user members; and
- informing user members on developments in telecommunication.

ATUG is a user body set up to take all reasonable action for users to obtain

- the availability of a wider range of telecommunication services and products; and
- satisfactory quality of telecommunication services and products at optimum prices.

ATUG is based on the conviction that competition in the telecommunication marketplace is essential. The primary activities of this user's group is political lobbying and presentations to organizations involved with legislation or policy for telecommunication in Australia. ATUG also consults with the major carriers, namely Telecom Australia, OTC, and Aussat and provides information to members.

ATUG provides submissions prepared by either the secretariat, task forces, or board committees. The group has a number of service activities including newsletters, publicity, conferences, special interest groups, provision of information, and state operations. A board of directors is elected annually, and there are membership fees for various categories of members.

11.15 STANDARDS ASSOCIATION OF NEW ZEALAND (SANZ)

> Standards Association of New Zealand
> Private Bag
> Wellington
> New Zealand
>
> Telephone: 64 4 84 2108
> Facsimile: 64 4 84 3938

The present Standards Association was created by the Standards Act of 1965, which provided for a large Standards Council, independent of governmental control, to develop New Zealand's national standards system. The Standards Act of 1988 reduced the Standards Council to fewer members, but it is still broadly representative of New Zealand organizations involved in standardization.

The Standards Act of 1988 requires the Standards Council to develop standards and to promote their use with the objective of improving the quality of goods and services, promoting standardization, encouraging industrial development and trade, and promoting safety, health, and welfare.

The Standards Association of New Zealand has a 65-person staff directly employed by the member body. The Standards Association's primary responsibilities include the preparation of standards, sale of publications, marking of goods, quality control services, education, promotion, and testing facilities.

11.16 TELECOM CORPORATION OF NEW ZEALAND, LTD.

Access Standards Section
Corporate Policy Department
Telecom Corporate Office
PO Box 570
Wellington
New Zealand

Telephone:	64 4 823 045
Facsimile:	64 4 851 702

Telecommunication in New Zealand was totally deregulated on 1 April 1989. There is no direct equivalent of AUSTEL in New Zealand, but its Ministry of Commerce oversees competitive activities and monitors progress on deregulation.

Each network operator has the right to set technical standards of a network. In this respect, Telecom has published specifications to advise suppliers on the requirements for gaining permission to connect their terminal equipment or for interconnecting their own networks to the Telecom network. So far, Telecom is the only fully established network operator and whether prospective network operators will adopt Telecom's standards or set their own is not known.

Telecom's standards are closely based on CCITT Recommendations, where applicable, but also align with other nations for specific services and applications. As an example, some of Telecom's specifications are closely aligned with the equivalent British standards.

The Telecom group formally responsible for the publication of interconnection standards is the Access Standards Section. These interconnection standards are based on engineering work within other specialized areas of

Telecom. The Access Standards Section is part of the Corporate Policy Department reporting to the company secretary. The corporation is a registered company wholly owned by the New Zealand government.

New Zealand Telecom's standards-setting activities are almost entirely internal, and are based on its own decisions regarding local implementation of CCITT Recommendations and overseas standards. Under arrangements agreed with the Ministry of Commerce, Telecom's specifications are issued in draft form for public comment. Special interest groups, such as ITANZ (Information Technology Association of New Zealand) and TUANZ (Telecom Users Association of New Zealand), and the local industry are given the opportunity to make any suggestions for improvements or changes.

The Standards Association of New Zealand (SANZ) is likely to take an increased interest in standards-setting activities once other network operators are established, as SANZ can adopt an independent coordinating role. So far, however, this work has been left almost entirely to Telecom, where much of the expertise currently resides.

11.17 INFORMATION TECHNOLOGY ASSOCIATION OF NEW ZEALAND (ITANZ)

Information Technology Association of New Zealand
Pilmmer City Centre, Level 30
Corner Boulcott Street and Gilmer Terrace
Wellington
New Zealand

Telephone: 64 4 722 731
Facsimile: 64 4 711 745

ITANZ is an association of suppliers of information technology products to New Zealand. The main objectives of the organization are to achieve a satisfactory environment within New Zealand for the information technology industry and to provide business opportunities for its members. The association represents the views of the industry to important influential organizations, such as government, state-owned enterprises, and Telecom. Issues addressed are current and include standards, telecommunication services, copyright, health, government procurement policies, and education.

Members of ITANZ seek to maximize their potential by associating with other members and developing "partnerships in opportunity." ITANZ members may also join in special interest groups and participate in task forces that help formulate industry views to influence their field of interest.

ITANZ has a president, past-president, vice president, and permanent executive director, and a ten-member executive board of chief executive officers of multinational and New Zealand companies, elected by members. Task forces are initiated through subcommittees to address particular issues. The task force findings are communicated via the executive board to appropriate organizations within New Zealand. Linkages for members are provided through newsletters, strategy groups, task forces, and special events.

11.18 SINGAPORE INSTITUTE OF STANDARDS AND INDUSTRIAL RESEARCH (SISIR)

Singapore Institute of Standards and Industrial Research
Kent Ridge
PO Box 1128
Singapore 9111

Telephone:	65 778 77 77
Facsimile:	65 778 00 86

Standardization activities in Singapore began in 1966, and in 1967 the Industrial Research Unit (IRU), a technical department under the Economic Development Board (EDB), Ministry of Finance, became a member of ISO. In 1969, the IRU was reorganized into the Singapore Institute of Standards and Industrial Research (SISIR), and standardization was officially incorporated as part of the institute's functions. In 1973, the institute was incorporated as an autonomous statutory body under the SISIR Act 1973. SISIR comes under the purview of the Ministry of Trade and Industry, and is administered by a board of management, which endorses the Singapore standards on the recommendation of a standards council. SISIR's main thrust is in quality and technology. In quality, SISIR promotes and helps companies implement the concept of companywide quality improvement. SISIR also provides support for certification, quality consultancy, expert technical assistance, laboratory accreditation, and metrology. In technology, SISIR supports the local industry in industrial research and development. Areas covered include product design and development, product testing and evaluation, food technology, material technology, and technology transfer.

The SISIR has a staff of 346 directly employed by the member body. SISIR's primary responsiblities include preparation of standards, sale of publications, marking of goods, quality control services, metrology, education in promotion, testing facilities, and applied industrial research.

11.19 THE STANDARDIZATION COUNCIL OF INDONESIA (DEWAN STANDARDISASI NATIONAL, DSN)

Standardization Council of Indonesia
Gedung PDII-LIPI
Jalan Gatot Subroto
PO Box 3123
Jakarta 12190
Indonesia

Telephone: 62 215 83465
Telex: 62875 pdii ia

The Standardization Council of Indonesia (DSN), a governmental department, was established by a presidential decree in 1984. It has a 39-person staff directly employed by the member body.

The DSN's primary objectives are to

- coordinate, synchronize, and maintain cooperation among institutions concerned with standardization activities; and
- submit advice and consideration to the President of the Republic of Indonesia concerning national standardization policy.

In particular the DSN is to

- formulate and determine the national policy on standardization;
- coordinate, synchronize, and evaluate the standardization program and activities;
- carry out international cooperation, and coordinate participation of the institutions concerned with various international organizations and bilateral, regional, and international technical cooperation on standardization;
- approve the concept of the consensus standards to be national standards;
- carry out the active solution of problems (if any) among the institutions concerned and be the information center for standardization; and
- formulate and approve the formulation procedures for national standards, implementation criteria, and other standardization activities.

Primary responsibilities of the DSN include preparation of standards, sale of publications, education, and promotion.

11.20 REPUBLIC OF KOREA BUREAU OF STANDARDS (KBS)

Republic of Korea Bureau of Standards
Industrial Advancement Administration
2, Chungang-Dong Kwachon-City
Kwonggi-do 171-11
South Korea

Telephone:	82 2 503 79 28
Facsimile:	82 2 503 79 41

The expansion of industrial civilization followed by the growth of industries in the Republic of Korea has brought about a new social climate in which the needs and benefits of mass production are strongly felt among manufacturers. The government of Korea responded to the need to foster industrial standardization and to improve the quality of domestic industrial products for export and domestic consumption. In 1961, the government promulgated the Industrial Standardization Law, which regulated the nationwide standardization activities in Korea. Simultaneously, the Bureau of Standards (KBS), responsible for the administration of industrial standardization, was set up under the Ministry of Commerce and Industry. KBS was hence reorganized under the Industrial Advancement Administration in the governmental reorganization of 16 January 1963.

The KBS, a member of ISO, has a 53-person staff directly employed by the member body and also a staff of 38 working for the member body, but sponsored by other organizations. The primary responsibilities of KBS include the preparation of standards, marking of goods, and metrology.

11.21 PHILIPPINES BUREAU OF PRODUCT STANDARDS (BPS)

Philippines Bureau of Product Standards
Third Floor
Trade and Industry Building
361 Senator Gil J. Puyat Avenue
Makati Manila 3117
The Philippines

Telephone:	632 818 5705
Facsimile:	632 851 166

A more formal approach to standardization was made in the Philippines with the establishment of the standardizing body in 1947, when the Division of Standards was created, under the Bureau of Commerce, Department of Commerce and Industry. In June 1964, after realizing the need for standardization, in line with industrial and economic development of the nation, Congress enacted Republic Act No. 4109, converting the Division of Standards into the Bureau of Standards.

In 1987, the bureau was renamed as the Bureau of Product Standards (BPS), under the Department of Trade and Industry. The BPS is a governmental body that promotes, implements, and coordinates standardization activities in the country and is a member of ISO. BPS has an 87-person staff directly employed by the member body. The primary responsibilities of BPS include the preparation of standards, sale of publications, marking of goods, quality control services, education, promotion, and testing facilities.

11.22 THAI INDUSTRIAL STANDARDS INSTITUTE (TISI)

Thai Industrial Standards Institute
Ministry of Industry
Rama VI Street
Bangkok 10400
Thailand

Telephone: 66 2 245 7802
Facsimile: 66 2 246 8826 Attention TISI

The Thai Industrial Standards Institute (TISI) was brought into being within the Ministry of Industry in February 1969. TISI's objective is to prepare national standards and promulgate them as a direct contribution to the national economy and industrial rationalization. Certification of products was begun in 1972, and the institute now administers both voluntary and compulsory certification schemes. TISI technical committees comprise specialists from many professions, and standards are formulated under the familiar "consensus-of-opinion" principle.

The government has been a firm supporter of standardization since the inception of TISI, a notable event being the procurement regulation requiring all government purchasing officers to order by reference to standards and show preference for certified products. TISI also gains support from other government departments, associations, and universities, particularly in testing for compliance with standards. TISI holds punitive powers under the Industrial Products Standards Act B.E. 2511 (1968) and subsequent ministerial regulations require the institute to take action against indigenous manufacturers and importers in the event of noncompliance with compulsory standards.

TISI has a staff of 528 directly employed by the member body. In addition, there is a staff of 5,766 working for the member body, but sponsored by other organizations. TISI's primary responsibilities include the preparation of standards, sale of publications, marking of goods, quality control services, education, promotion, and testing facilities.

11.23 STANDARDS AND INDUSTRIAL RESEARCH INSTITUTE OF MALAYSIA (SIRIM)

Standards and Industrial Research Institute of Malaysia
PO Box 35, Shah Alam
Selangor
Malaysia

Telephone:	635 559 2601
Facsimile:	603 550 8095

SIRIM was established in 1975 with the passing of Act 157, known as the Standards and Industrial Research Institute of Malaysia (Incorporation) Act, 1975. SIRIM was formed as a result of the merger of the Standards Institution of Malaysia (SIM) and the National Institute for Scientific and Industrial Research (NISIR), to provide for the preparation and promotion of standards for commodities, processes, and practices, the promotion and undertaking of industrial research, and matters related to those purposes.

Principally, SIRIM is to

- promote, develop, and promulgate standards for commerce and industry and for goods produced in or imported into Malaysia, whether for consumption in Malaysia or for export or re-export;
- promote industrial efficiency and development;
- promote public and industrial welfare, health, and safety;
- promote and undertake industrial research with the objectives of
  —improving technical processes and methods,
  —discovering new processes and methods,
  —encouraging the utilization of Malaysian products, and
  —adapting technology developed in other countries for use in Malaysia;

- apply the results of research;
- provide consultative services to assist industry in meeting standards; and
- improve production processes and techniques.

SIRIM, a member of ISO, has a staff of 714 directly employed by the member body. SIRIM's primary responsibilities include the preparation of standards, sale of publications, marking of goods, quality control services, metrology, education, promotion, testing facilities, and applied industrial research.

11.24 CHINA STATE BUREAU OF STANDARDS (CSBS)

China State Bureau of Standards
China State Bureau of Technical Supervision
PO Box 2112
Beijing
People's Republic of China

| Telephone: | 861 444 304 |
| Facsimile: | 861 895 098 or 861 401 1016 |

The governmental body for standardization in the People's Republic of China was set up in 1957, as the State Scientific and Technological Commission and Bureau of Standards. In 1972, the body became an independent government department titled the State Bureau of Standards and Metrology. In 1978, the latter was divided into two organizations, one of which being the China State Bureau of Standards (CSBS).

The main functions of the CSBS are the

- elaboration of the state principles and policies on standardization;
- organization of the development and provision of national standards;
- approval issuance and publication of those standards; and
- administration of the supervision and inspection of product quality as well as the provision of standards information.

CSBS participates in international standards activities on behalf of the Chinese government, and is a member of ISO. CSBS has a staff of 665 directly employed by the member body. In addition, a staff of 21,911 works for the member body, but is sponsored by other organizations. The primary responsibilities of the CSBS include the preparation of standards, sale of publications, marking of goods, quality control services, education, and promotion.

CHAPTER 12

CONFORMANCE TESTING, FUNCTIONAL STANDARDIZATION, AND INTERNATIONAL STANDARDIZED PROFILES

12.1 INTRODUCTION

This chapter first discusses functional standards and international standardized profiles. The three regional bodies involved in functional standardization and standardized profiles, for Europe (EWOS), Asia and the Pacific (AOW), and North America (OIW), are then considered. The functional standardization organizations involved in conformance testing in each of the three regions contribute to the regional bodies. These organizations are mainly SPAG in Europe, POSI in Japan, and COS in the United States.

We then consider some of the application organizations, which have been working in the area of manufacturing and office systems. Throughout we present a regional point of view.

12.2 FUNCTIONAL STANDARDIZATION IN ISO-IEC JTC 1

The Meaning of Functional Standardization

Work on OSI started in March 1978, "to enable interconnection and interworking of computer systems of different makes by means of data communication." OSI is constructed in seven layers that are independent of each other. For each of the seven layers, several standards have been established or are under preparation, dealing with the services in each layer and their protocols. There are now well over 75 ISO standards (or draft standards), some of which consist of several parts. These base standards offer a broad functionality by means of a large number of "degrees of freedom." Such degrees consist of parameters, options, subsets, classes, and so on. The details were not specified in the standards, however, because they would then have been too long in the making, particularly as consensus would have been harder to achieve.

In terms of the OSI objective, the base standards constitute a *necessary condition for* achieving interworking (or interoperability). To achieve a specific user function, however, electronic mail, for instance, one needs a combination of base standards in a coherent set with a resolution of all the degrees of freedom. Computers from different vendors will only interwork if the same set of standards and resolutions of freedom within such standards are implemented on the different machines. This gives rise to the concept of a *profile*.

A profile defines a working set of one or more base standards necessary to provide a specific function or set of functions. The profile standardizes the use of particular options available in the base standard. A profile can never change the base standard.

A *functional standard* is a document describing a profile. A functional standard also contains conformance statements, which indicate capabilities selected for implementation in a real system and list static and dynamic behavioral properties of the implementation.

Functional standards are created and finally published by ISO-IEC JTC 1 as *international standardized profiles* (ISPs). An ISP is "an internationally agreed to, harmonized document which identifies a group of (one or more) standards, together with options and parameters, necessary to accomplish a function or set of functions."

ISO-IEC JTC 1 decided to establish a Special Group on Functional Standardization (SG-FS) to set up and maintain the procedure to establish ISPs. The SG-FS has developed the Technical Report Type 3 ISO/IEC/TR 10 000.

Document TR 10 000 describes

- the purpose of profiles,
- the concept of an ISP,
- the taxonomy of ISPs,
- the structure of documentation for profiles,
- a directory of profiles and ISPs, and
- ISO-IEC drafting rules adapted for ISPs.

The directory contains the classification of profiles in *transport* (T), *application (A)*, and *interchange format* (F) layers (or levels) as well as an index of these profiles. The index will be continuously updated due to its dynamic character and contains information giving the ISP in which a profile is documented.

Workshops and the Complementary Process

Functional standards can be prepared, ratified, and published wholly or partly within JTC 1. The majority, which are prepared and ratified outside JTC 1, are done so in "the complementary process." The complementary parts consist of two activities:

Part I

1. The technical definition and creation process;
2. The global harmonization of the result of that process.

Part II

3. The public ratification process (review and approval);
4. The publication as an international standardized profile.

Generally, Part I is done outside JTC 1 in regional Workshops on Open Systems (WOSs), and Part II is done within JTC 1. Part I is regulated by bylaws of the regional workshops and Part II by the JTC 1 procedures. The three regional workshops that have been established are

- the National Institute of Standards and Technology (NIST) Workshop for OSI Implementors, for North America;
- the European Workshop for Open Systems (EWOS), for Western Europe; and
- the OSI Asia Oceania Workshop (AOW), for Eastern Asia and Oceania.

The workshops are fully backed by vendor and user organizations active in functional standardization and related activities, such as conformance or interoperability testing and demonstrator projects. These organizations are

- Corporation for Open Systems (COS), in the United States;
- World Federation of MAP/TOP Users, in the United States (including EMUG and OSITOP in Europe);
- Promoting Conference for OSI (POSI), in Japan; and
- Standards Promotion and Application Group (SPAG), in Europe.

The four organizations COS, MAP/TOP, POSI, and SPAG cooperate in the *Feeders' Forum*. Within this forum they deliver a major contribution to the second activity in Part I of the functional standardization, the global harmonization.

The importance of these vendor and user organizations has been recognized by ISO and IEC in their offering such groups liaison status (S-liaison) and designating them as "OSI User Groups." Through the S-liaison, the OSI User Groups are authorized to participate in the functional standardization work within JTC 1 in the same way as that allowed to the permanent members (P-membership) of the SG-FS. Voting rights in JTC 1, however, are restricted to P-members.

The flow of information from regional workshops to SG-FS through its members consists of

- proposals for profiles for incorporation in the directory of profiles and ISPs;
- announcement of draft ISPs;
- proposed draft ISPs (pdISPs);
- explanatory report attached to each pdISP; and
- maintaining information on and updates to existing ISPs.

The informational items listed above support the smooth transition between Part I in the regional workshops and Part II in JTC 1. This information prevents JTC 1 from being taken by surprise and ensures a full-fledged standards-making process in the regional workshops, where all parties concerned can participate. In this way, technical consensus can be achieved and controversies can be kept outside Part II in JTC 1.

Table 12.1 charts the flow of work in JTC 1. The stages 3 through 5 (review, vote, comments, resolution, and publication) take from seven to ten and one-half months. This is fast compared to any comparable procedure in ISO-IEC JTC 1, and is known as the "turbo procedure."

The Work Program

The work program on draft ISPs has a dynamic character. There is a larger number of profiles and corresponding ISPs defined and included in the directory than those described below, which are nonetheless representative. Four pdISPs were announced in 1987 and submitted to JTC 1 in 1989 (the sponsoring organization is given in brackets):

Table 12.1. Procedure for Publishing ISPs

Activity	Phase	Stage	Time (months)	Actors
proposal for profile in Directory	inclusion in Directory	1	<6	Taxonomy Group + SG-FS
update of Directory	pre-development	?	?	external to SG-FS
announcement of pdISP	1 development 2 preparation of review	2	?	1 external to SG-FS 2 SG-FS
submission of pdISP + Explanatory Report + Appointment of editor	review	3	1 to 2	Editor + Review Team + SG-FS + ISO Central Secretariat
distribution of dISP for voting	voting	4	<3 (+1)	P-members JTC1
voting results and comments become available	resolution of comments	5	1 1/2 to 2	Editor + Review Team + SG-FS
submission of ISP for publication	publication	6	<1 1/2 +1	Central Secretariat
publication of ISP	maintenance	7	?	external to SG-FS

- packet-switched data network (PSDN), permanent access;
  - —connection-oriented transport service (COTS) over connection-oriented network service (CONS) [POSI];
  - —COTS over connectionless network service (CLNS) [COS];
- CSMA/CD local area network (LAN): COTS over CLNS; [MAP/TOP];
- file transfer, access and management (FTAM); simple file transfer [SPAG].

ISPs being considered in 1990 are

- token bus LAN: COTS over CLNS;
- message handling system (MHS), public domain access, user agent (UA), and message transfer agent (MTA);
- MHS, private domain access, UA, and MTA; and
- office document format.

12.3 FEEDER'S FORUM

JTC 1 develops ISPs, and will accept proposed draft ISPs from a number of groups working on OSI standards. Coordination of the international groups is necessary , however, so that regional differences in functional standards are settled before a pdISP is submitted to JTC 1.

The mechanism for harmonization is the Feeder's Forum, which comprises the NIST workshop on open systems, the Asia-Oceania workshop for open systems, and the European workshop on open systems, which are discussed below.

The Feeder's Forum is organized on two levels:

- Management Level Feeder's Forum (MLFF), which coordinates and guides activities, and has formal liaison with ISO and IEC; and
- Technical Level Feeder's Forum (TLFF), which coordinates and carries out technical work.

The secretariat of the Feeder's Forum changes each year, being, for example, with Japan in 1989. The third MLFF and the seventh TLFF were held in Belgium, in July 1989.

12.4 OSI ASIA-OCEANIA WORKSHOP (AOW)

Interoperability Technology Association
for Information Processing (INTAP)
OSI Asia-Oceania Workshop
Sumitomo Gaien Building 3F
24, Daikyo-cho
Shinjuku-ku, Tokyo 160
Japan

Telephone:	81 3 358 27 21
Facsimile:	81 3 358 47 53

OSI Asia-Oceania workshop was formed on 19 October 1988 to assist in developing internationally harmonized implementation specifications for OSI. AOW provides a forum for OSI experts in the Asia-Oceania region in the development of pdISPs. The AOW is to ensure acceptable interoperability based on OSI. The AOW also works to promulgate and promote the use of ISPs. AOW is structured to reflect adequately the views of the people from the countries concerned. AOW conducts workshops on pdISPs and coordinates views in cooperation with similar workshops in North America and Europe, and contributes the results to ISO-IEC JTC 1.

Organization

AOW consists of a council, a plenary meeting, and individual special interest groups (SIGs). The council comprises organizations wishing to promote interoperability of information processing in countries of the Asia and Oceania area. These organizations are registered at the secretariat and are approved by the existing council. The plenary meeting and the SIGs are composed of workshop members, who are researchers or engineers of research organizations, standardization organizations, computer vendors, and users of information processing technology.

The annual plenary meeting elaborates project plans, establishes, disbands, and coordinates SIGs, and approves results of discussions by SIGs. The SIGs deliberate and discuss pdISPs in accordance with directions of the plenary meeting.

Adoption of a resolution at the plenary meeting, in principle, is based on unanimous approval of all participants. Any opposition requires presentation of reasons and an alternative proposal. If unanimous approval cannot be reached, the matter is subjected to voting by members present and a resolution is passed by a majority of two-thirds or more.

The initial council members are the China State Bureau of Technical Supervision (People's Republic of China), Electronics and Telecommunications Research Institute (Republic of Korea), Interoperability Technology Association for Information Processing (Japan), and Standards Australia (Australia). There are SIGs on WAN, ODA, LAN, FTAM, MHS, and the directory. AOW participates in coordinating meetings with the EWOS and NIST OSI workshops.

At the first coordinating meeting of AOW, EWOS, and NIST, on 6 March 1989, working methods and other topics were discussed. AOW was charged with developing a pdISP for WAN as the originating organization, and submitted a final version to ISO-IEC JTC 1, SG-FS, in 1989.

SIG Activities

File Transfer Access and Management SIG (FTAM SIG)

The File Transfer Access and Management SIG (FTAM SIG) has approved the profile for *unstructured file transfer with the character attached* as an addendum called AFT11. FTAM SIG is now considering other profiles such as AFT12 (flat file transfer), AFT22 (flat file access), and AFT3 (file management), and playing a coordinating role for the character set among the workshops.

Office Document Architecture SIG (ODA SIG)

ODA is the name of International Standard ISO 8613, which defines interchange format of office documents. The application layer standard is being developed in ISO-IEC JTC 1, SC 18.

ODA functional standards developed in Europe and the United States have been introduced, and drafts of three international core profiles (i.e., CORE 11, CORE 26, and CORE 36) have been reviewed, and there has been a discussion on language and culturally dependent features, such as character sets.

Coordination among the various groups involved in ODA functional standards has been through a liaison group formed from the OSI groups, informally called *profile alignment group on ODA* (PAGODA).

CORE 11 supports a simple document structure and character content; CORE 36 supports an advanced document structure and character content, raster graphics content, and geometric graphics content; and CORE 26 has the functionality of CORE 11 and CORE 36 and supports three kinds of content.

Wide Area Network SIG (WAN Lower Layer SIG)

Ratification of the final pdISP has been completed with discussions of new profiles to be agreed by the regional workshop coordinating committee. The full title of this pdISP is Information Processing System–International Standardized Profiles TB, TC, TD, TE–Connection-Mode Transport Service over Connection-Mode Network Service.

Local Area Network SIG (LAN SIG)

A pdISP is being discussed as of 1990. This is one of the T-profiles (OSI lower layer profiles) and provides connection-mode transport services over connection-mode network services.

Two types of subnetworks are now included in the scope of this profile (TA 51: CSMA/CD; TA 111: packet-switched data network). The scope will need to be extended to include other types of subnetworks (e.g., token-passing bus, token-ring).

12.5 EUROPEAN WORKSHOP FOR OPEN SYSTEMS (EWOS)

European Workshop for Open Systems
Second Floor
13, rue Brederode
B-1000 Brussels
Belgium

Telephone: 32 2 511 7455
Facsimile: 32 2 511 8723

The European Workshop for Open Systems was created in December 1987 by the most representative European federations of technology suppliers and user organizations: Cooperation for Open Systems Interconnection Networking in Europe (COSINE), European Computer Manufacturers Association (ECMA), European MAP Users Group (EMUG), Open Systems Interconnection Technical and Office Protocols (OSITOP), Réseaux Associés pour la Recherche Européene (RARE), and the Standards Promotion and Application Group (SPAG), in conjunction with the European standards institutions, CEN and CENELEC. The foregoing are also the members of EWOS's steering committee, chaired by H. Donner (SPAG).

From the beginning, the Commission of the European Communities has supported this initiative. Meanwhile, DG IX has become a member of EWOS's steering committee.

EWOS's objectives are to serve as a truly open European forum for the development of OSI profiles and definition of the corresponding conformance testing specifications.

All results of the EWOS's activities are fed into the formal standardization channels managed by the European and international standardization bodies, ISO-IEC JTC 1 and CEN-CENELEC.

Working Procedures

The EWOS structure is small and flexible, and can therefore easily adapt to changing circumstances. The development process, however, is monitored as an industry-like project in which the deadline for results is set at the start. These results are contained in formal proposals, called EWOS documents (EDs). These serve as direct proposals to the standards bodies for adoption as, for example, European prestandards (ENVs) or standards (ENs). Other useful information not meant for standardization is issued as an EWOS technical guide (ETG).

EDs are formally approved and numbered documents, which are prestandards proposals for a

- functional standard (or profile);
- conversion of an existing ENV into an EN, which is issued as a "maintenance statement";
- specification for the conformance testing of a corresponding profile; or
- a liaison statement to a pdISP.

ETGs are also formally adopted and numbered documents, which relate to the prestandardization work of EWOS and aim at facilitating the correct selection and use, implementation, and coherent further development of additonal standards or prestandards. The ETG may be a

- new taxonomy proposal (to ITAEGS, ITAEGM) within the framework of the information technology memoranda (and later also TR 10 000);
- tutorial document (especially for new areas or a revised approach; or
- set of guidelines for simplifying implementation of the functional standard in information technology products.

The technical activities of EWOS are managed by a technical assembly (TA), chaired by U. Hartmann (Siemens).

Some 120 experts from 79 European and US companies have become TA members. As expected, these members comprise users, suppliers, standardization institutes, government agencies, and academics. Seven TA meetings were held during 1988 and 1989. The timetable for 1990 contained four further EWOS workshops and TAs.

Program of Activities

The current program of EWOS activities covers seven domains of which virtual terminal, directory, and manufacturing message specification were recently added. Five national member bodies offer technical secretarial support to the expert groups.

The work underway at present and the responsible members are given by Table 12.2.

EWOS has adopted a General Policy for work on conformance test specification standards. A study and investigation mandate has been issued by the Commission of the European Communities, after positive advice from the competent national representatives, to support common action by CEN-CENELEC via EWOS and ETSI, the recognition arrangements, OSTC and ETCOM, and the coordination bodies, ITSTC and ECITC.

An expert group on conformance testing has been established by the last TA to formulate proposals regarding the execution of the mandate and further work of EWOS in this area. A number of proposals for new work items have been considered by the EWOS steering committee.

For most of these work items, the CEC has proposed mandates, formally adopted by the relevant bodies. The proposed topics are

Table 12.2. Current EWOS Expert Group Activities (*circa* 1990)

Domain	EWOS/EG	Member Body
Directory	DIR	NNI
File Transfer Access and Management	FTAM	NNI
Lower Layers	LL	DIN
Message Handling Systems	MHS	AFNOR
Manufacturing Messaging Services	MMS	DIN
Office Document Architecture	ODA	IBN
Virtual Terminal	VT	BSI

- Study of a General Framework for the Elaboration of IT Functional Standards;
- X/Open Portability Guide;
- Use of OSI in the Libraries' Community;
- Application of OSI Based Solutions in the Medical Field; and
- Remote Access to Databases.

These topics are being examined by the EWOS TA. The study and investigation mandate of the CEC should cover the first feasibility studies and suggest corresponding action plans.

Liaison and Coordination

Liaison has been established with the North American OSI Implementors' Workshop and the Asia-Oceania Workshop for Open Systems. Since March 6, 1989, a Regional Workshop Coordinating Committee (RWS-CC) has been established to synchronize the work programs of the RWs on a more permanent basis, and a second meeting of this committee was held in Washington, in 1989.

EWOS accepted the editorship and responsibility for submission to JTC 1 of 10 of the 18 pdISPs accepted at the 1989 meeting. To process contributions to pdISPs, EWOS has obtained S-liaison status to ISO-IEC JTC 1 SG-FS. EWOS also has close liaison with ETSI on documents and meetings on common topics.

12.6 NIST OSI IMPLEMENTORS WORKSHOP

National Institute of Standards and Technology
National Computer Systems Laboratory
Gaithersburg, Maryland 20899

Telephone: 301 975 3608
Facsimile: 301 975 2128

The NIST OSI Implementors Workshop (OIW) was set up in January 1988 to develop agreements based on emerging OSI standards. Participation and commitment to OIW is voluntary, and there is no membership as such.

The OIW has an executive committee and a plenum, both of which meet four times a year. OIW has 14 special interest groups, which are

- Directory Services,
- FTAM,
- Lower Layers,
- Manufacturing Message Specification,
- Network Management,
- ODA,
- OSINET TC,
- Registration,
- Remote Database Access,
- Security,
- Transaction Processing,
- Upper Layers,
- Virtual Terminal, and
- X.400.

There are 12 persons on the executive committee, which is elected by the workshop plenary assembly. The plenary assembly comprises all SIG workshop delegates and meets to approve the SIGs' recommendations.

OIW has liaison with EWOS, with a view to submitting pdISPs to ISO-IEC JTC 1 for approval. OIW operates under a procedures manual which can be amended by any plenary assembly.

12.7 STANDARDS PROMOTION AND APPLICATION GROUP (SPAG)

SPAG Services, S.A.
149, avenue Louise, Box 7
B-1050 Brussels
Belgium

| Telephone: | 32 2 535 0811 |
| Facsimile: | 32 2 537 2440 |

The Standards Promotion and Application Group (SPAG) was founded in 1983 as a consortium of European information technology companies. The Commission of the European Communities was instrumental in the creation of SPAG to coordinate activities within Europe toward the development of European functional standards. SPAG became a company registered under Belgian law as SPAG Services, S.A., in 1986.

Each SPAG shareholding company has an equal number of shares and the board of directors currently has eleven members. Six seats on the board are held by original founding companies: Bull, ICL, Nixdorf, Olivetti, Philips, Siemens. The other five seats are held by Alcatel, British Telecom, Digital Equipment Corporation, Hewlett Packard, and IBM, all of whom have since joined.

SPAG's mission is to achieve a truly open international market for the computer and telecommunication industries, based on harmonized standards, testing, accreditation, and certification for OSI products. To achieve this aim, SPAG plays a significant role in a number of European and worldwide activities related to the achievement of multivendor interworking based on international standards protocols.

In the area of functional standards, SPAG plays a leading role in EWOS, has participated in the development of ISPs through the Feeder's Forum, and publishes the Guide to the Use of Standards (GUS).

SPAG is active in the development and marketing of conformance testing tools and interoperability testing services (ITS) for office and manufacturing protocols. Through its involvement in the recognition arrangement, European Testing and Certification for Office and Manufacturing (ETCOM), SPAG is helping to establish an accreditation and certification infrastructure for LAN-based OSI products within Europe.

In addition, SPAG is actively involved in a wide range of CEC research projects. The results from ESPRIT Project 955 Phase 3 (CNMA-CCT), which developed conformance testing tools, are now being commercially exploited through the SPAG-CCT consortium in Europe. (The ITS is also available through a distributor network in North America and Japan.) Due to this success, ESPRIT Project 2292 TT-CNMA has been initiated for which SPAG is the prime contractor. The objectives of the project include the promotion and development of IOP testing, performance measurement, conformance testing, and integration of related technology and other CEC programs. SPAG is also actively involved in two CTS2 projects: one focusing on FTAM, where SPAG is developing executable test cases and embedded testing; the other dedicated to X.400 MHS and X.500 DS, whereby SPAG is concentrating on DS test development and harmonizing related work within the TT-CNMA project. In the area of telecommunication development and future networks, SPAG heads a consortium to harmonize testing and testing technology and to develop verification methodologies and tools for integrated broadband communication networks within RACE (under the current PROVE project).

SPAG has also established close working relationships with its counterparts in North America (COS) and Japan (POSI), which led to the establishment of the C-P-S Forum.

12.8 CORPORATION FOR OPEN SYSTEMS (COS)

Corporation for Open Systems
1750 Old Meadow Road
Suite 400
McLean, Virginia 22102-4306

Telephone:	703 883 2700
Facsimile:	703 848 4572

Insofar as every manufacturer was previously working on its own version of an OSI standard communication system, many of the individual versions would have been incompatible with one another. COS provides a means of pooling resources so that work is not duplicated and incompatability is resolved.

COS was set up in North America as a result of initiative in 1985 by the Computer and Communication Industry Association (CCIA). The stated objective of COS is similar to that of SPAG, "to accelerate open systems architecture implementation by supplementing existing organization channels." Approximately 40 of the largest North American companies in computers and telecommunication became founding members. There is a central secretariat comparable to that of the CCITT and a budget of approximately $10 million. In addition to involvement in standards-setting, COS has facilities for the testing and certification of members' systems for conformance with the group's own specifications.

Among the objectives of COS are to

- accelerate development of standards that are key to making OSI a reality in high-priority applications;
- provide technical input to the standards recommendations that the CCITT study groups develop;
- support and participate in the development of OSI standards by ISO; and
- work actively on the US technical committees X3, X9, and X12, and on the T1 committee of the Exchange Carriers Association, IEEE, EIA, and other standards-developing organizations.

International Standards and Implementation Agreements

International standards can only be used as a basis for protocol selection if those standards already exist in a sufficiently mature form. If a COS objective is being delayed by an insufficient degree of international consensus or there is no current external activity relating to a COS priority, COS works for the creation of such an activity within a suitable forum or establishes one within the group's own structure.

Testing Facilities and Certification

Testing facilities are used for ensuring conformance specifications and interoperability of products built to those specifications. COS is developing an integrated testing architecture to enable

- tests for conformance and interoperability;
- testing support for the different stages of the product life cycle;
- transfer of testing technology; and
- an authoritative product certification scheme.

Certification is needed to give some authority to the results of a formal product testing process. A respected and authoritative certification mark helps users in considering a purchase and vendors in the marketing of their products. To ensure that vendors can sell in a variety of geographical markets, worldwide harmonization of certification schemes will be necessary.

Functional Profiles

COS is working toward

- accreditation of functional profiles, test facilities (suites, systems, and centers);
- changing the control of functional profiles and test facilities to represent the stages of development of interoperability circles;
- development and accreditation of certification criteria; and
- development of definitions and procedures for limitations of liability when legal remedies are sought.

Once selected by a functional standards group, a profile must be validated for operability, a function usually implicit in the interoperability requirement, and registered with qualified testing centers. Test suites provide the working definitions for functional profiles, and must be approved by the functional standards group whose protocols are purportedly validated. Tests must be performed in accordance with developed and accredited certification criteria, approved by some interest group or global forum so that the testing tools and results accord with universal quality.

12.9 PROMOTING CONFERENCE FOR OPEN SYSTEMS INTERCONNECTION (POSI)

Promoting Conference for Open Systems Interconnection
Japan Electronic Industry Development Association
(JEIDA)
Kikai-Shinko Kaikan
3-5-8, Shiba Koen 3-chome
Minato-ku, Tokyo 105
Japan

| Telephone: | 81 3 433 19 41 |
| Facsimile: | 81 3 433 63 50 |

The Promoting Conference for Open Systems Interconnection (POSI) was established in November 1985, in response to the Japanese government's policy of international cooperation and to promote open systems interconnection. The founders and current members of POSI are Oki Electric Industry Co. Ltd., Toshiba Corp., NEC, Hitachi Ltd., Fujitsu Ltd., Mitsubishi Electric Corp., and Nippon Telegraph and Telephone Corporation (NTT). The first six are members of POSI, and NTT and Japan Electronic Industry Development Association (JEIDA) are observers.

POSI has a board of directors, an executive committee, a steering committee, and a technical committee. Under the technical committee, there is one special working group and three expert groups:

- Conformance SWG,
- FTAM EG,
- WAN EG, and
- ODA/ODIF EG.

POSI has been instrumental in establishing Feeder Forums with other functional standards bodies, COS, SPAG, and MAP/TOP.

POSI has also been instrumental in forming AOW for the specification and support of international standardized profiles and in supporting the AOW-EWOS-NIST Regional Workshop Coordination Committee (RWCC). In 1989, POSI took part in a strategic technical cooperation meeting (named the CPS meeting) with COS and SPAG in Tokyo, regarding an agreement for the harmonization of testing technology.

12.10 WORLD FEDERATION OF MAP/TOP USERS GROUPS

In January 1986, the Society of Manufacturing Engineers (SME) was appointed secretariat to the World Federation of MAP/TOP Users Groups. In January 1989, Industry Cooperatives Services, Inc., assumed the secretariat role.

The World Federation of MAP/TOP Users Groups is a formal organization of international user's groups dedicated to the adoption and use of internationally accepted MAP (Manufacturing Automation Protocol) and TOP (Technical and Office Protocol) specifications, based on international standards. This organization comprises

- North American MAP/TOP Users Group,
- European MAP Users Group,
- Japan MAP Users Group, and
- Australian MAP/TOP Interest Group.

Membership will include the East European MAP/TOP Interest Group (EEMIG).

The objectives of the MAP/TOP Users Groups are to

- promote a full appreciation of the advances of MAP/TOP for users and vendors;
- harmonize international efforts into a single coherent specification document;
- encourage and support the rapid development of international standards related to these systems such that MAP/TOP can be totally configured on international standards and test mechanisms;
- ensure that MAP/TOP hardware and software products are available in a universal format;
- generate a standard set of communication specifications using cur-

rent industry standards, user-driven MAP and TOP specifications, and accepted industry operating practices;

- improve productivity by implementing a common communication specification referenced in a variety of office equipment, manufacturing hardware, and computer-based systems;
- promote interoperability of multivendor devices to improve the integration of engineering, office, and manufacturing devices;
- use the OSI model to assist in the definition of standards and involve MAP/TOP Users Groups to accelerate acceptance; and
- provide market feedback to computer and other device manufacturers, encouraging the development and support of nonproprietary communication products that implement common specifications.

MAP

MAP provides the common set of communication protocols for multivendor factory floor environments. MAP is compatible with TOP and is based on the seven-layer OSI model.

The North American MAP/TOP Users Group shares a common steering committee, which has eleven user companies and nine *ex-officio* members. The MAP executive committee comprises vice-chairs representing the areas of programming and product, membership and training, standards organizations and associations, and technical and test.

A number of special interest groups have been formed, including processing industry, aerospace, government, business case, and fiber optic. Recently, the North American MAP/TOP Users Group merged with COS.

12.11 EUROPEAN MAP USERS GROUP (EMUG)

European MAP Users Group Secretariat
College of Manufacturing, Building 30
Cranfield Institute of Technology
Cranfield, Bedford MK43 0AL
England

| Telephone: | 44 234 752 794 |
| Facsimile: | 44 234 750 882 |

EMUG was set up early in 1986. It is a regional user's group for Europe, working in close collaboration with the North American MAP/TOP Users Group and the corresponding groups in Australia and Japan. Together they form the World Federation of MAP/TOP User Groups, working to ensure acceptance of the worldwide data communication standard, MAP, which is a realization of the seven-layer OSI model. EMUG is now devoted to the encouragement of computer-integrated manufacturing (CIM) in Europe.

EMUG has over 150 corporate members from the EEC, EFTA, and Eastern European (COMECON) countries. The members are approximately equally divided among users and vendors with a portion of observer members. EMUG also has national chapters in the United Kingdom, France, Denmark, Norway, and Sweden to service small companies.

EMUG is overseen by a general manager with secretariat staff. All other positions are voluntary and unpaid. Once a year (via an annual general meeting) a steering committee of up to a maximum of 20 members is elected from the paid membership. The steering committee comprises user and vendor representatives at a ratio of 3:1. The steering committee may appoint its own subcommittees (e.g., for marketing and promotional activity). The officers are selected from among the steering committee members and are a chair, one or two vice-chairs, and a treasurer. Other committees are open to all members. The technical committee has three related groups: implementation and application, technical work item, and architecture.

The objectives of EMUG are to

- unite European efforts toward an international MAP format;
- contribute to the rapid development of international bases and functional standards for MAP and CIM;
- inform and promote understanding of MAP in Europe;
- encourage, support, and promote MAP pilot and demonstration projects in Europe;
- ensure MAP production availability in Europe;
- establish independent European MAP conformance testing and certification;
- act as a unified voice of European users to pressure vendors of computer and CIM equipment to deliver MAP products;
- improve European knowledge of MAP and CIM through cooperative technical work;
- represent European views internationally in the MAP community through the World Federation of MAP/TOP User Groups;
- recognize competence centers to provide members with necessary

skills and training to assist their installation and use of MAP; and
- provide migration strategies from existing networks to MAP.

EMUG continually works to achieve these objectives via different activities. EMUG supported the Enterprise Networking Event (ENE88) in Baltimore, demonstrating MAP 3.0 prototypes, and also arranged the parallel MAP/TOP/OSI Symposium in England. EMUG also has sponsored a special MAP exhibition at SYSTEC in Munich and cooperates with other organizations in providing exhibits or presentations at related conferences.

12.12 INTERNATIONAL ROBOTICS AND FACTORY AUTOMATION CENTER (IROFA)

> International Robotics and Factory Automation Center
> Daiichi Nakano Building, Ninth Floor
> 2-6-10, Iwamoto-cho
> Chiyoda-ku, Tokyo 101
> Japan
>
> Telephone: 81 3 861 5601
> Facsimile: 81 3 861 5635

Japan's center for research and development in robotics and automation (IROFA) is a nonprofit organization, founded through the approval of the Ministry of International Trade and Industry (MITI) in June 1985. IROFA performs research and development in cooperation with industrial, academic, national, and public research organizations (including research laboratories of MITI) concerned with factory automation (FA). At present, the IROFA members comprise 110 companies.

The organizational structure of IROFA includes of a chair, vice-chair, senior executive director, executive director, and secretariat. Under the secretariat is a planning office and five departments (research, technical, business, international, and general affairs. To each of these are attached research, technical, business, international, and general affairs divisions, respectively. An accountants' division is also attached to the general affairs department.

IROFA is involved in the following activities concerning standards-setting:

- ISO/TC 184,

- TC 184/SC 5, and
- research study on standardization for safety and reliability of factory automation systems.

The first project was committed to IROFA in 1986 for internal deliberation in Japan, concerning ISO/TC 184 (Industrial Automation System) and ISO/TC 184/SC 5 (System Integration and Communication) by the Standards Department of the Agency of Industrial Science and Technology in MITI.

IROFA has a committee for the standardization of factory automation, which consists of three technical committees (system interface, cell interface, and information interface) and one subcommittee (programming language).

12.13 OSIone

OSIone is a single cooperative association comprising five regional OSI network groups: EurOSInet (Europe), INTAPNET (Japan), OSIcom (Australia), OSINET (US), and OSNET (Singapore). The objective of OSIone is to undertake matters of common interest in support of the establishment of a unified platform for regional OSI networks to facilitate the promotion and advancement of worldwide OSI connectivity and interoperability.

OSIone enables regional OSI network groups to cooperate globally while retaining autonomy in their respective groups. The group promotes OSI through demonstrations and other activities in support of OSIone's purpose and mission. OSIone uses the following as the basis for demonstration and promotion to maximize interoperability:

- internationally harmonized functional standards wherever possible, or
- appropriate regional functional standards.

Through collaborative activities, OSIone's members identify the differences in regional OSI profiles and provide guidance on interoperability and demonstrate the maximum level of service available to international OSI users. OSIone also assists in global convergence of profiles from otherwise divergent regional implementations.

OSIone has an executive committee comprising one representative of each member organization. This committee meets in plenary sessions twice a year. Other committees are organized as required to manage the administrative and technical needs of the group.

The initial focus of OSIone's demonstrations at various locations is the needs of the international users of X.400 message handling services, but will be progressively enhanced to cover the full spectrum of OSI applications.

12.14 OSIcom

> OTC Limited
> 231 Elizabeth Street
> Sydney NSW 2000
> Australia
>
> Telephone: 61 2 287 5318
> Facsimile: 61 2 287 4122

OSIcom is a group committed to the implementation of OSI standards. The group is collaborating to establish a network of systems called OSIcom, which will be available for the demonstration of multivendor interworking applications using OSI standards. OSIcom is intended to have an Australian orientation, but eventually linking with like overseas organizations.

The object of OSIcom is to offer a continually available demonstration of interworking between information systems using agreed OSI application profiles and emerging standards. This approach will

- raise the level of market awareness and confidence of OSI as a practical solution to problems of systems interworking;
- raise Australian skill levels in OSI; and
- facilitate the development of OSI products in Australia.

OSIcom is managed by a committee consisting of four officers and four ordinary members of the association. Chairs of subcommittees are entitled to be a members of the committee. The four officer positions are chair, vice-chair, treasurer, and secretary. There are currently three technical subcommittees, one each for X.400, FTAM, and OSI management plus one user's subcommittee.

Subgroups are required to develop appropriate demonstrations intended for inclusion in the group's portfolio. The portfolio includes, for each demonstration, a list of members available to participate and an outline technical description. The precise definition of a demonstration profile is determined by the committee, based on recommendations of the appropriate sub-group, and takes account of relevant and agreed functional standards. All members of OSIcom are not required to participate in those demonstrations.

12.15 OPEN SYSTEMS INTERCONNECTION NETWORK (OSINET)

National Institute of Standards and Technology (NIST)
Technology Building 225, Room B217
Gaithersburg, Maryland 20899

Telephone: 301 975 3631
Facsimile: 301 975 2128

OSINET is an international organization established in 1984 to foster the development, promotion, and deployment of OSI through activities related to interoperability testing. OSINET is engaged in three different areas of activity:

- research and development of test suits and the conduct of OSI inter-operability testing;
- demonstration and promotion of the OSI technology; and
- testing and registration of announced OSI products.

OSINET is an organization of potential users and suppliers of OSI products. Members include industry user companies, government user agencies, and supply companies. OSINET is sponsored by NIST, but is governed by the group's membership. OSINET consists of one physical and two organizational entities

- OSINET network, comprised of subnetworks, intermediate systems, and end systems;
- OSINET Steering Committee; and
- OSINET Technical Committee.

The responsibility of the steering committee is to establish and manage OSINET. The steering committee determines and approves all OSINET projects. The responsibility of the technical committee is to carry out the technical program of work.

Organizations satisfy membership requirements either through participation in a research and development project, in a testing and registration project, or in both. To maintain eligibility in OSINET, an organization must attend three of the most recent four meetings of the OIW (see Section 12.6).

12.16 THE OSITOP EUROPEAN USERS GROUP (OSITOP)

OSITOP European Users Group
21, avenue de Messine
F-75008 Paris
France

Telephone:	33 1 4042 6643
Facsimile:	33 1 4042 2579

Formation and Objectives

OSITOP was formed in February 1987 as a nonprofit user's association in Belgium, founding members being BNP, Bull, CCTA, CIIBA, DISTRIGAZ, EDF, GDF, Hoechst, ICL, KEMA, NCC, and Siemens. OSITOP's objectives are to

- study and develop standardization of products conforming to standards in engineering, office automation, and communication;
- encourage and support the development of international standards in accordance with OSI;
- promote a worldwide consensus on the development of functional standards;
- affect specific European requirements within engineering, office automation, and communication standardization bodies;
- promote the establishment of testing and certification centers in Europe; and
- harmonize and coordinate with European and international bodies to ensure a uniform approach.

In summary, OSITOP's objectives are to

- include the European user's requirements in the ongoing standardization process;
- promote ISO standards and international standardized profiles; and
- set up pilot projects for its members.

Organization

OSITOP members are either participating users, participating vendors, or observers, who may be either users or vendors. OSITOP has a

- steering committee,
- technical committee, and
- secretariat.

The steering committee, which is the governing body of OSITOP, has up to fifteen user member representatives and up to five vendor member representatives elected by the general meeting. User members must be in the majority by at least 3:1 over vendor members. The steering committee handles all management and administration, subject to powers allocated to the general meeting.

Guidelines for the technical committee, its organisation, and structure are the responsibility of the steering committee, which presents a working program for approval by the annual general meeting. The technical committee oversees the activities delegated to working groups with a view toward the agreed program.

The technical committee harmonizes the efforts of working groups and acts as liaison with PTTs in Europe, ITSTC, CEN, CENELEC, CEPT, SPAG, ECMA, EMUG, ETSI, and ISO. OSITOP is an S-Liaison member of ISO, a founding member of EWOS, and an observing member of ETSI.

Technical Work

OSITOP's technical activity is through working groups, which are

WG1:	Standards Analysis and Evaluation which provides information on the ongoing standardization process, and gives guidance to users);
WG2,4:	Products, Projects and Migration (which are involved in developing pilot projects and studying migration issues);
WG3:	Testing and Certification Group; and
WG5:	Architecture Group (which has responsibility for definition of OSITOP architecture and the relationship between functional profiles for technical and general office environments).

The OSITOP technical committee has produced the following documents:

- Present Status of Testing and Certification,
- Aspects of Current Standardization and Future Developments,
- FTAM, and
- X.400.

The X.400 project (see Table 12.3) has been installed and operated free of charge by Bull in France and Hewlett Packard in Germany.

12.17 GOVERNMENT OPEN SYSTEMS INTERCONNECTION PROFILES (GOSIP)

In 1988, the British government's Central Computer and Telecommunications Agency (CCTA), which was the first to recognize formally the increasing power of procurement policy in having suppliers work toward open systems, published its Government OSI Profile (GOSIP). It sets specifications for OSI-based products used within the government to ensure that separately purchased and different systems can work together. The US government has its own set of GOSIP standards.

The first UK document was know as Version 3.0 to position it alongside MAP/TOP Version 3.0 profile. GOSIP Version 3.1 was published in February 1990, updating the 1988 version of X.400 message handling, including directory, security, electronic data interchange, and distribution elements.

UK GOSIP, however, is more than a list of standards in that it serves as a menu to users and a guide to suppliers. The CCTA divides the seven-layer OSI model into two parts:

- GOSIP-T deals with the transport stack, and
- GOSIP-A with the application stacks above the transport layers.

GOSIP was primarily devised for administrative systems in government departments. GOSIP, however, need not be restricted to purely governmental use, and is not written with government solely in mind.

As useful as it has been in the United Kingdom, GOSIP has still greater international potential. In addition to the US version, a number of countries,

Table 12.4. North American ISDN User's Forum Organizational Structure

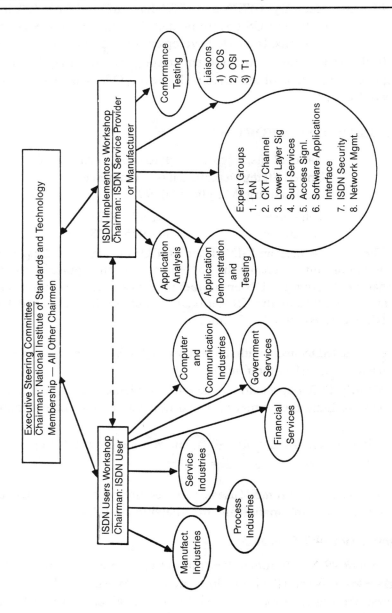

- application requirements,
- application profiles,
- implementation agreements, and
- conformance criteria.

The IUW produces application requirements, which describe potential applications of ISDN and the features that may be required. The IIW develops application profiles, implementation agreements, and conformance criteria, which provide the detailed technical decisions necessary to implement an application requirement in an interoperable manner. The activities within the two workshops are coordinated by the NIU Forum's executive steering committee. Further particulars on the workings of the NIU Forum are in Appendix 12A. This appendix includes a description of work on application profiles and the application profile teams.

APPENDIX 12A

THE OPERATION OF THE NIU FORUM

ISDN User's Workshop (IUW)

The IUW is responsible for identifying, defining, and setting priorities for user requirements as well as working with the IIW to reach agreements to support the implementation of user requirements. Activities within the IUW are coordinated by its steering committee. Membership of the IUW is open, and there are seven industry membership groups.

Any user can request consideration of a particular ISDN application, which is a solution to a business problem and is based on cost reduction, productivity enhancement, application interfaces, and performance improvement.

User applications are first submitted to the NIU Forum through the IUW industry groups. The sponsoring industry group develops a user application request into a formal *application requirement* document and identifies a point of contact for the application. The user concerned works closely with the IIW Applications Analysis Working Group to ensure that the final document contains all necessary information. Each application is evaluated by the appropriate industry group and is then forwarded to the steering committee.

An approved and coordinated user application requirement is assigned an identification number by NIST and sent to the users with the minutes of the meeting. Users review the application and vote at the next plenary session to determine the priority of the application requirement, which is used by the IIW to allocate resources. After priority is set, application requirements are submitted through the NIU executive steering committee to the IIW for processing.

ISDN Implementor's Workshop (IIW)

The IIW is responsible for developing *application profiles, implementation agreements,* and *conformance criteria* to support IUW defined application requirements, providing technical advice and consultation to the IUW, multivendor demonstrations and trials, and formal liasion with appropriate organizations, such as COS or the T1 Committee. Membership of the IIW is open to any organization, and all members may vote, provided that they have attended two consecutive meetings.

The IIW is organized into working groups (WGs), which include an Applications Analysis WG, Application Demonstration and Promotion WG, Conformance Testing WG, application profile teams, and a number of expert WGs. The Applications Analysis WG conducts initial analysis of user application requirements and develops application analyses to serve as a baseline for the development of application profiles. The Conformance Testing WG is responsible for defining requirements and providing guidelines for interpreting the results of conformance tests. The Application Demonstration and Testing WG coordinates the demonstration of multivendor implementations of application profiles. The expert WGs develop implementation agreements and provide expertise to the application profile teams. Expert WGs include access signaling, local area networks, supplementary services, software-applications interface, security, and terminal adaption. Application profile teams develop profiles from user requirements. The IIW steering committee is responsible for coordinating the working groups' activities.

Application Profiles

The application profile contains the recommended set of implementation agreements for all seven layers of the OSI model, which must be present to support a particular user's application or set of applications. An application profile is prepared by a team using an application analysis developed by the Application Analysis WG. This WG serves as the liaison between the IUW and IIW, evaluates the application requirement, and works with the team to resolve questions about the application. Users are encouraged to participate with the application profile teams in the development of application profiles. This practice is especially beneficial if the user who originates the suggestion actively participates in the process.

The Application Analysis WG concentrates on understanding the business function that must be satisfied and, as appropriate, documents optional technical solutions. The result is an applications analysis that translates the user's application requirement into a document that can be used by the IIW to develop the necessary application profile. This document is presented to the ITW plenum for consideration, coordination, and to provide an opportunity for the entire membership to comment on other options. Changes are made as appropriate and the results of the completed application analysis are presented to the IUW for approval.

Once the IUW approves an application analysis, the IIW steering committee assigns the application to an application profile team. This team will develop an application profile from the user's application requirement and the applications analysis. Application profile team members will include representatives from both the appropriate expert groups and users.

The IIW currently has six application profile teams. The teams, which handle families of applications, are

- Network Management/ISDN Administration,
- Call Management,
- Network Interconnectivity,
- Messaging and Answering,
- Bandwidth Negotiation, and
- CPE Compatibility/Capability.

Whenever possible, the application profile also references existing implementation agreements (IAs) which have been approved by the NIU Forum's executive steering committee. If the implementation agreement does not exist, the team will coordinate with the appropriate expert group to develop the IA or a draft IA to facilitate consensus. This procedure ensures that IAs are developed to satisfy user application requirements.

Implementation Agreements

Implementation agreements define protocols in terms of the specific options and parameters of standards and recommendations that must be present to satisfy a particular application. The objective of the IA is to ensure interoperability, and only those options which affect interoperability need to be specified.

Many IAs have already been developed within the OSI Implementor's Workshop (OIW), which is another NIST-sponsored industry forum. Approved IAs are published by NIST in reports titled, "Stable Implementation Agreements for ISDN Protocols" or "Stable Implementation Agreements for Open Systems Interconnection Protocols." These published agreements will be considered for adoption by NIST as new or revised federal information processing standards, which can be used to establish clear purchasing standards. The IAs published by the NIU Forum are limited to standards developed by appropriate industry forums: CCITT, ISO, IEEE, EIA, ANSI, and NIST. New IAs are developed by the expert groups as needed to satisfy user application requirements.

The NIU Forum does not prepare standards, but influences industry standards forums as appropriate. If a standard does not exist, a liaison statement will be prepared and forwarded to the appropriate forum.

Conformance Criteria

Conformance criteria contain test scenarios and procedures used to certify products that satisfy implementation agreements contained in a particular version of ISDN. Conformance criteria are prepared by the IIW Conformance Testing WG. The actual conformance tests are implemented and administered by organizations such as COS, which are specifically set up for this purpose. The NIU Forum is not a testing and certification organization.

Industry Coordination

The NIU Forum's executive steering committee is responsible for coordinating the activities of the forum with those of other organizations. The forum is concerned with both national and international progress in ISDN-related activities.

OSI Implementor's Workshop

Some of the expert support needed by the IIW to process user application requirements into application profiles is provided by the OIW, also sponsored by NIST. Both the Network Management and Packet Expert WGs of the IIW have been consolidated with existing Special Interest Groups in the OIW. In addition, the OIW has agreed to use the IUW as the mechanism by which ISDN user application requirements can be defined and introduced into the OSI workshop. NIST coordinates activities between the two forums to counter duplication of effort. A long-term objective of NIST is to integrate OSI and ISDN technologies.

Corporation for Open Systems

The NIU Forum's executive steering committee has invited COS to participate in the forum to coordinate the specification and development of conformance criteria for testing and certification. COS aims toward worldwide information system interoperability, and develops and conducts tests by which information system products can be evaluated for interoperability with respect to OSI and ISDN.

ISDN Versions

The IUW benefits from the direct involvement of the OIW and COS in the NIU Forum. Work in the MAP/TOP Users Group has made apparent that versions of ISDNs must be treated as they exist at a certain point in time, derived from existing national and international standards; further, each version should be compatible with earlier ones. Manufacturers and service providers are expected to develop ISDN offerings based on a particular version. One could then associate it with an event that demonstrates interoperable applications of the version. This practice was used successfully by the MAP/TOP Users Group in the Enterprise 88 event in Baltimore, in June 1988, to promote OSI.

The Application Demonstration and Testing WG is planning a similar event to promote the strategic development of ISDN in North America. This event would demonstrate the availability and interoperability of products that use an agreed version of ISDN.

Since the NIU Forum's inception in June 1988, over 300 companies have participated and more than 60 applications have been presented to the IUW for consideration.

IIW application profile teams are currently working on several different user application requirements. The development of application profiles is a new undertaking, and the teams are developing recommended operating procedures in addition to their work on application profiles. The Call Management Application Profile Team has taken a leadership role in developing these procedures. This team is developing an application profile to allow incoming calling line identification (CLID) to access automatically a host application for customer information and to display this information on an agent terminal when the call is answered. This profile would also pass critical customer information to a second agent should the call be transferred. This particular profile includes sections on the scope of the application, information flows, implementation options, and the seven-layer protocol suite required to implement the application.

Although the focus of the IUW is on identifying and defining applications, the IUW also accepts *problem resolution requests* (issues) from the participants. These requests are coordinated through the industry group, the IUW steering committee, and the IUW plenum. If required, the problem resolution request is forwarded to the executive steering committee for action. Several issues have been identified that may require a problem resolution request for the executive steering committee. Some users are concerned about the possible legal issues associated with the ISDN CLID. This feature enables a called party to identify a calling party's telephone number. Users who prefer to have an unlisted telephone number may not want this feature used on their calls. Another issue which has been raised in the IUW is user concern about costs associated with ISDN. Users would want the service providers to develop models to identify the cost elements. Some users have also expressed concern about the ISDN numbering plan. The problem resolution request provides users with a formal method for addressing issues of concern.

CHAPTER 13

OTHER INTERNATIONAL ORGANIZATIONS

13.1 INTRODUCTION

This chapter deals with organizations that are broadly international in character. Some have a philosophical and social approach as well as a technical approach and are more hybrid in that they consider the general effect of computing and information processing in addition to telecommunication. Such groups are the International Chamber of Commerce (ICC),* IIC, and IFIP. The major group that focuses user interests, INTUG, is considered, and reference is made to a number of international organizations concerned with standards.

13.2 INTERNATIONAL TELECOMMUNICATIONS USERS GROUP (INTUG)

International Telecommunications Users Group
Secretariat
18 Westminster Palace Gardens
Artillery Road
London SWIP IRR
England

Telephone: 44 71 799 2446
Facsimile: 44 71 799 2445

INTUG was created in 1974 to be a collective voice of telecommunication user associations and to represent its members' interests in international telecommunication affairs.

---

* The ICC, although it fits the description of such an organization, is not treated in detail here.—Ed.

INTUG represents national telecommunication user associations from Australia, Austria, Canada, Finland, Hong Kong, Japan, New Zealand, Switzerland, the United States, and Western Europe.

INTUG's objectives are to ensure that users' views are heard and allowed to contribute to the adoption of new technology facilities and the formulation of tariffs and standards. INTUG seeks to achieve its goals by creating and maintaining continuous cooperation with telecommunication authorities and encouraging dialogue with national and international bodies concerned with telecommunication. INTUG's mandate is to pursue the introduction of new facilities, oppose excessive regulation, and encourage rapid adoption of new technology. INTUG encourages efficient management of telecommunication facilities so that users will benefit from cost savings and improved services.

More specifically INTUG's constitution states that the objective of the association is to promote the interests of all telecommunication users on an international basis. To further this objective, INTUG

- fosters the development of telecommunication policies, services, standards, and regulations best suited to user needs, at prices attractive to users;
- pursues, organizes, and participates in studies to determine present and future user needs;
- promotes cooperation among national user groups;
- organizes and develops dialogue between users and other national and international bodies, such as ITU, CEPT, EEC, OECD, PTTs; and
- provides an international mechanism to inform users on matters affecting telecommunication.

Membership

Membership is open to any person or group acceptable to the council and subscribing to the objectives of INTUG, except public telecommunication network providers. Companies whose business is concerned with a supply of telecommunication products are to be represented by staff concerned as users rather than as providers. There are three grades of membership: a full member, which is a national or international user association with a formal constitution conforming to the objectives of INTUG; an associate member, which is an organization with an interest in international telecommunication as a user rather than as a provider; and individual members, who may be academics, researchers, students, *et cetera,* interested in telecommunication.

The council of INTUG comprises the officers and representatives of full members. The officers consist of a president, chair of the council, treasurer, and such other officers deemed necessary by the council. Vice-chairs are appointed to administer the geographic regions of Europe, Asia and the Pacific, and the Americas. The officers of INTUG form the executive committee.

There is provision for annual, general, extraordinary, and plenary meetings, which are to be held not less than once each calendar year and primarily devoted to the work program and related subjects. Executive and council meetings of INTUG have the power to establish committees on special subjects.

At the moment two significant working groups are: an INTUG working group on liaison effort with the CEC and a working group liaising with EUSIDIC on the performance of international public data networks.

INTUG works in liaison with the ICC, CEC, and the Organization for Economic Cooperation and Development (OECD). INTUG participates in the CCITT, primarily in the areas of service guidelines and tariff principles.

INTUG represents the interests of a range of users, both small and large, and is effective not only because it understands the PTTs' environmental constraints and points of view, but because it follows through with appropriate action. INTUG's approach is to engage in constructive dialogue rather than to take strong partisan positions. INTUG provides an excellent conduit through which users can make their views known.

13.3 INTERNATIONAL INSTITUTE OF COMMUNICATIONS (IIC)

International Institute of Communications
Tavistock House South
Tavistock Square
London WC1 H9LF
England

Telephone: 44 1 388 06 71
Facsimile: 44 1 380 06 23

Purpose of the IIC

The International Institute of Communications is an independent, non-governmental, nonprofit organization, dedicated to a better understanding of communication issues throughout the world.

The IIC is involved in communication research and policy analysis, particularly on telecommunication and broadcasting issues. The IIC reflects the global interdependence of communication across a variety of sectors and disciplines. The IIC provides an informal, flexible, free, and dispassionate forum for discussion and debate by all people concerned with the future of communication. The institute's mission is to understand communication in the broader sense and to evaluate the finer points that politics, business, and culture bring to bear on the use of communication around the world.

The IIC has been engaged in research on telecommunication issues since 1969, when its predecessor, the International Broadcasting Institute (IBI), was founded after meetings in the United States, Europe, and Japan. In response to the expanding interests of its membership and the convergence of distribution technology, the IBI became the IIC in 1977. Its focus now includes satellites, switched digital communication, cable television, and video.

Structure

Membership is open to all individuals and organizations who are concerned about the future of communication. Organizations may join as corporate (profit-making) or institutional (nonprofit) members.

There are active independent national associations of the IIC in Canada, France, the Federal Republic of Germany, Italy, Japan, the United Kingdom, Australia, and the United States. Each national association reflects the culture and makeup of the communication sector.

Management

The IIC is governed by a board of trustees which is elected by the membership each year at the annual conference. The board of trustees elects the officers, who form the executive committee and are responsible for the daily management of the institute in close collaboration with the directorate and staff at the IIC's London office.

Annual Conference

The annual conference brings together IIC members and provides an opportunity for politicians, industrialists, and academics from developed and developing countries to meet and discuss key issues in communication. Those participating in the conference may present their views as individuals rather than as representatives of their discipline or organization. The 1989 conference was held in Paris, and the 1990 conference was held in Dublin.

Research and Projects

The IIC carries out a large number of policy analyses and research studies. These projects are primarily for the IIC's members, although other organizations and individuals may be invited to participate. Past projects include

- a study of telecommunication structures;
- a study of television production;
- a series of activities on the use of telecommunication for economic development;
- a report for the government of Japan on multinational alliances in the telecommunication industry; and
- a review of advanced high definition television systems.

Current projects include

- a study and debating forum on the evolution of national telecommunication policies; and
- a study of ownership and control of television in the 1990s.

13.4 INTERNATIONAL FEDERATION FOR INFORMATION PROCESSING (IFIP)

International Federation for Information Processing
16, place Longemalle
CH-1204 Genève
Switzerland

Telephone: 41 22 28 2649
Facsimile: 41 22 781 2322

The International Federation for Information Processing (IFIP) is a multinational aggregation of professional and technical societies (or groups of such societies) concerned with information processing. Only one such society or group in any country can be admitted as a full member, which must be representative of the national activities in the information processing field. At present, IFIP has approximately 46 national societies as members, representing 64 countries.

The aims of IFIP are to

- promote information science and technology;
- advance international cooperation in the field of information processing;
- stimulate research, development, and application of information processing in science and human activity;
- further dissemination and exchange of information on information processing; and
- encourage education in information processing.

IFIP was officially established in 1960. The supreme authority of IFIP is the general assembly, which meets annually. The executive body of IFIP is the president, three vice-presidents, secretary, and treasurer, who are elected by the general assembly. The daily work of IFIP is administered by a secretariat.

The council, consisting of the officers and up to eight elected trustees, meets twice a year and makes decisions between general assembly meetings.

IFIP has established a number of technical committees (TCs) and working groups (WGs). The current technical committees are

TC2	Programming
TC3	Education
TC5	Computer Applications in Technology
TC6	Data Communication
TC7	System Modeling and Optimization
TC8	Information Systems
TC9	Relationship between Computers and Society
TC10	Digital Systems Design
TC11	Security and Protection in Information Processing Systems
TC12	Artificial Intelligence
TC13	Human-Computer Interaction

Altogether, these technical committees have fifty-four working groups. TC6 (Data Communication) has four working groups, including

WG6.4   Local Computer Networks

WG6.5   International Computer Message Systems

13.5 INTERNATIONAL COUNCIL FOR COMPUTER COMMUNICATION (ICCC)

International Council for Computer Communication
PO Box 9745
Washington, DC 20016

Telephone: 301 530 76 28

The International Council for Computer Communication (ICCC), founded in 1972, is a nonprofit corporation registered in the District of Columbia.

ICCC's members, called governors, are limited to 100, and elected for terms of six years by governors in office. They are drawn from all disciplines and fields of endeavor related to computer communication (e.g., scientists, technologists, equipment and system developers, service providers, users, economists, lawyers, and government officials).

The ICCC is an affiliate member of the IFIP. The ICCC was formed to foster

- scientific research and development of computer communication;
- progress in evaluation of applications of computer communication to educational, scientific, medical, economic, legal, cultural, and other peaceful purposes; and
- study of potential social and economic effects of computer communication and policies that influence those effects.

The ICCC has concentrated on providing stimuli and means for communication between people of many interests and disciplines, who could contribute to the general understanding of potential advances, risks, problems, and solutions related to computer communication.

At first, the emphasis was on technical issues of data network design, transmission and switching techniques, control, and compatibility. The thrust of the ICCC's work was to have these issues resolved in the context of real needs, projected usage, and economics. Currently, there is an effort to address opportunities unique to developing regions of the world. More recently, the implications on computer communication of developments in personal computers and artificial intelligence are being considered.

The principle activity of the ICCC is sponsorship of biennial International Conferences on Computer Communication, forums for presentation of viewpoints, projection of needs, debate of issues, and reporting of plans and accomplishments. In 1990, the conference was held in New Delhi.

13.6 REGIONAL GROUPINGS OF POST AND TELEPHONE ADMINISTRATIONS

The major non-European regional groupings of PTT administrations are

- African Postal and Telecommunication Union (UAPT)
- Arab Telecommunication Union (ATU)
- Asia-Pacific Telecommunication Union (APT)
- Inter-American Telecommunication Union (CITEL)
- Pan-African Telecommunication Union (PATU).

In addition, CEPT has been considered in Chapter 10, APT has been considered in Chapter 11, and, by way of illustration, COPANT is considered below.

13.7 PAN-AMERICAN STANDARDS COMMISSION (COPANT)

Pan-American Standards Commission
Avida Julio A. Roca 651
Piso 3, Sector 10 (1322)
Buenos Aires
Argentina

The Pan-American Standards Commission (COPANT) coordinates at a regional level the standardization work of 15 institutes, one of which, ICAITI, comprises the five Central American republics. COPANT collaborates in the development and promotion of standardization at the national level as well.

COPANT has approved 1542 standards, and is now studying 300 documents developed by technical study committees. The commission is studying the possibility of implementating a seal of conformity with COPANT standards.

13.8 OTHER INTERNATIONAL ORGANIZATIONS

International Maritime Satellite Organization
(INMARSAT)
40 Melton Street
London NW1 2EQ
England

Telephone: 44 1 387 9089
Facsimile: 44 1 387 2115

International Society for Aeronautical Telecommunications
(SITA)
112, avenue Charles de Gaulle
Neuilly-sur-Seine
F-92522 CEDEX
France

Telephone: 33 1 473 85000
Facsimile: 33 1 474 75606

Society for Worldwide Interbank Financial
Telecommunications S.C.(SWIFT)
1, avenue Adèle
B-1310 La Hulpe
Belgium

Telephone: 32 2 655 3111
Facsimile: 32 2 655 3226

International Air Transport Association (IATA)
2000 Peel Street
Montréal, Québec H3A 2R4
Canada

Telephone: 514 844 6311
Facsimile: 514 844 5286

International Civil Aviation Organisation (ICAO)
Place de l'Aviation Internationale
1000 Sherbrooke Street West
Montréal, Québec H3A 2R2
Canada

Telephone: 514 285 8221
Facsimile: 514 288 4772

International Radio and Television Organization (OIRT)
Technical Center: Skokanska
16956 Praha 6
Czechoslovakia

International Telecommunications Satellite Organization
(INTELSAT)
3400 International Drive NW
Washington, DC 20008-3098

Telephone: 202 835 6000 (Headquarters)
Telephone: 202 944 6800 (Australian Office)

Interim European Telecommunications Satellite
Organization (EUTELSAT)
Tour Maine Montparnasse
33, avenue du Maine
F-75755 Paris CEDEX 15
France

Telephone: 33 1 453 84747
Facsimile: 33 1 453 83700

International Maritime Organization (IMO)
4 Albert Embankment
London SE1 7SR
England

Telephone: 44 1 735 7611
Facsimile: 44 1 587 3210

International Press Telecommunication Council (IPTC)
Studio House, 184 Fleet Street
London EC4A 2DU
England

Telephone: 44 1 405 2608
Facsimile: 44 1 404 4527

European Association of Information Services (EUSIDIC)
First Floor Offices, 9/9a High Street
Calne, Wilshire SN11 OBS
England

Telephone: 44 0 249 813 584
Facsimile: 44 0 249 813 656

Glossary

ABTT	ISO-IEC Advisory Board on Technical Trends
ACEC	Advisory Committee on Electromagnetic Compatibility
ACES	Advisory Committee on Safety
ACET	Advisory Committee on Electronics and Telecommunications
AE	Application Elements
AFNOR	Association Française de Normalisation
AG	JCT 1 Subcommittee: Advisory Group
AI	Artificial Intelligence
AIIA	Australian Information Industry Association
ANSI	American National Standards Institute
AOW	Asia-Oceania Workshop
API	American Petroleum Institute
APT	Asia-Pacific Telecommunity
ASTM	American Society for Testing and Materials
ATU	Arab Telecommunication Union
AUSTEL	Australian Telecommunications Authority
AUSTEL-SAC	Australian Standards Committee
B-ISDN	Broadband ISDN
BCTS	Broadcasting and Communications Technology Satellite
BDT	Telecommunications Development Bureau
BEITA	Business Equipment Information Technology Association
BPS	Philippines Bureau of Product Standards
BSI	British Standards Institute
BSS	Broadcasting Satellite Service
CASCO	ISO Council Committee on Conformity Assessment
CASCO WG	CASCO Working Group
CATV	Community Antenna (Cable) Television
CB	Citizens Band
CBEMA	Computer and Business Equipment Manufacturers' Association
CCA	CENELEC Certification Agreement
CCI	International Consultative Committee of the ITU
CCIA	Computer and Communication Industry Association
CCIR	International Radio Consultative Committee
CCITT	International Telegraph and Telephone Consultative Committee
CCTA	Central Computer and Telecommunications Agency

CD	Committee Draft
CECC	CENELEC Electronic Components Committee
CEN	European Committee for Standardization
CEN-CENELEC	Joint European Standardization Organization
CENELEC	European Committee for Electrotechnical Standardization
CEPT	Conférence Européenne des Administrations des Postes et des Télécommunications
CERTICO	ISO Council Committee on Certification
CII	Center for the Informatization of Industry
CIM	Computer-Integrated Manufacturing
CISPR	Comité International Spécial des Perturbations Radioélectriques
CITEL	Inter-American Telecommunication Union
CLID	Calling Line Identification
CLNS	Connectionless Network Service
COMSOC	IEEE Communications Society
CONS	Connection-Oriented Network Service
COPANT	Pan-American Standards Commission
COPOLCO	ISO Council Committee on Consumer Policy
COS	Corporation for Open Systems
COSINE	Cooperation for Open Systems Interconnection Networking in Europe
COST	European Cooperation in the Field of Scientific and Technological Research
COTS	Connection-Oriented Transport Service
CSBS	China State Bureau of Standards
CTD	Center for Telecommunications Development
DBS	Direct Broadcast Satellite
DEVCO	ISO Council Development Committee
DIN	Deutsches Institut für Normung e.V.
DIR	Directory
DIS	Draft International Standard
DP	Draft Proposal
DSN	Standardization Council of Indonesia
EAB	ESPRIT Advisory Board
EBU	European Broadcasting Union
EC	European Community
ECMA	European Computer Manufacturers Association
ECOSOC	United Nations Economic and Social Council
ECQAC	Electronic Components Quality Assurance Committee
ECREEA	European Conference of Radio and Electronic Equipment Associations

ECSA	Exchange Carriers Standards Association
ECSC	European Coal and Steel Community
ECTEL	European Telecommunications and Professional Electronics Industries Association
ECTUA	European Council of Telecommunications Users Association
ED	EWOS Document
EDB	Economic Development Board
EDI	Electronic Data Interchange
EDRTS	Experimental Data Relay Tracking Satellite
EEA	Electronic and Business Equipment Association
EEC	European Economic Community
EEMIG	East European MAP/TOP Interest Group
EFTA	European Free Trade Association
EIA	Electronic Industries Association
EM	Equipment and Media
EMC	ESPRIT Management Committee
EMUG	European MAP Users Group
EN	European Standard
ENV	European Prestandard
ES	European Standards
ESCAP	United Nations Economic and Social Commission for Asia and the Pacific
ESPRIT	European Strategic Program for Research and Development in Information Technology
ETCOM	European Testing and Certification for Office and Manufacturing
ETG	EWOS Technical Guide
ETGC	Exchange Telephone Group Committee
ETS	European Telecommunication Standards
ETSI	European Telecommunications Standards Institute
EUCATEL	European Conference of Associations of Telecommunication Industries
Euratom	European Atomic Energy Community
EUSIDIC	European Association of Information Services
EUTELSAT	Interim European Telecommunications Satellite Organization
EWOS	European Workshop for Open Systems
FA	Factory Automation
FCC	Federal Communications Commission
FD	Formal Description
FIPS	Federal Information Processing Standards
FIPS PUBS	FIPS Standard Publications
FM	Factory Mutual Research Corporation

FM	Frequency Modulation
FRG	Federal Republic of Germany
FS	Fixed Services
FSS	Fixed-Satellite Service
FTAM	File Transfer Access and Management
GAP	Group for Analysis and Forecasting
GAS	Special Autonomous Group (*see* SAG)
GATT	General Agreement on Tariffs and Trade
GDP	Gross Domestic Product
GMDSS	Global Maritime Distress and Safety System
GOSIP	Government Open Systems Interconnection Profile
GSLB	Groupe Spécial Large Bande
GSO	Geostationary Satellite Orbit
GUS	Guide to the Use of Standards
HATS	Harmonization of Advanced Telecommunication Systems
HD	Harmonization Document
HDTV	High Definition Television
HFBC-87	Second Session of the World Administrative Radio Conference for the Planning of the High Frequency (HF) Bands Allocated to the Broadcasting Service, Geneva, 1987
I-ETS	Interim European Telecommunication Standards
IA	Implementation Agreement
IATA	International Air Transport Association
IBC	Integrated Broadband Communication
IBI	International Broadcasting Institute
ICA	International Communications Association
ICAO	International Civil Aviation Organization
ICC	International Chamber of Commerce
ICCC	International Council for Computer Communication
IDC	International Digital Communications, Inc.
IEC	International Electro-Chemical Commission
IEC	International Electrotechnical Commission
IEC GPC	IEC General Policy Committee
IECEE	IEC System for Conformity Testing to Standards for Safety of Electrical Equipment
IECQ	IEC Quality Assessment System for Electronic Components
IEEE	Institute of Electrical and Electronics Engineers
IFIP	International Federation for Information Processing
IFRB	International Frequency Registration Board
IIC	International Institute of Communications

IIIC	International Information Industry Congress
IIW	ISDN Implementor's Workshop
ILAC	International Laboratory Accreditation Committee
IMO	International Maritime Organization
IN	Intelligent Network
INFCO	ISO Council Committee on Information
INMARSAT	International Maritime Satellite Organization
INTAP	Interoperability Technology Association for Information Processing
INTELSAT	International Telecommunication Satellite Organization
INTUG	International Telecommunication User's Group
IPRC	Intellectual Property Rights Committee
IPS	Information Processing System
IPTC	International Press Telecommunication Council
IROFA	International Robotics and Factory Automation Center
IRU	Industrial Research Unit
ISM	ISDN Standards Management
ISO	International Organization for Standardization
ISONET	ISO Information Network
ISP	International Standard Profile
ITAC	Information Technology Association of Canada
ITANZ	Information Technology Association of New Zealand
ITJ	International Telecom Japan, Inc.
ITMG	Information Technology Management Group
ITS	Interoperability Testing Services
ITSTC	Joint Information Technology Steering Committee
ITTF	Information Technology Task Force
ITU-COM	Electronic Media Exhibition
IUW	ISDN User's Workshop
IWP	Interim Working Party
JEIDA	Japan Electronic Industry Development Association
JIPDEC	Japanese Information Processing Development Center
JIS	Japanese Industrial Standards
JISC	Japanese Industrial Standards Committee
JSA	Japanese Standards Association
JTC 1	ISO-IEC Joint Technical Committee on Information Technology
JTPC	ISO-IEC Joint Technical Programming Committee
KBS	Republic of Korea Bureau of Standards
KDD	Kokusai Denshin Denwa Co. Ltd.

LAN	Local Area Network
LL	Lower Levels
LPTV	Low Power Television
MAP	Manufacturing Automation Protocol
MAP/TOP	Manufacturing Automation Protocol/Technical and Office Protocol
MEDARABTEL	Regional Telecommunication Network in the Mediterranean and Middle East
MHS	Message Handling Service
MITI	Japanese Ministry of International Trade and Industry
MLFF	Management Level Feeder's Forum
MMS	Manufacturing Messaging Service
MOB-83	World Administrative Radio Conference for the Mobile Services, Geneva, 1983
MOSS	Market-Oriented Sector Specific
MOU	Memorandum of Understanding
MTA	Message Transfer Agent
MTN	Multilateral Trade Negotiation
NCSL	National Computer Systems Laboratory
NETS	Technical Recommendations
NHK	Japan Broadcasting Corporation
NISIR	National Institute for Scientific and Industrial Research
NISO	National Information Standards Organization
NIST	National Institute of Standards and Technology
NIU	North American ISDN Users
NPG	National Preparatory Group
NSG	National Study Groups
NTIA	National Telecommunications and Information Administration
NTT	Nippon Telegraph and Telephone Corporation
NWI	New Work Item
OBS	Office and Business Systems
ODA	Office Document Architecture
OECD	Organization for Economic Cooperation and Development
OIRT	International Radio and Television Organization
OIW	OSI Implementors Workshop
ONH	National Authorized Institution
ONP	Open Network Provision
ORB-85	First Session of the World Administrative Radio Conference on the Use of the Geostationary-Satellite Orbit and the Planning of the Space Services Utilizing It, Geneva, 1985

the Use of the Geostationary-Satellite Orbit and the Planning of the Space Services Utilizing It, Geneva, 1988

OSI	Open Systems Interconnection
OSINET	Open Systems Interconnection Network
OSITOP	European Group on Technical Office Protocols
PA	Plenary Assembly
PAGODA	Profile Alignment Group on ODA
PANAFTEL	Regional Telecommunication Network in Latin America, Asia, and Africa
PAR	Standards Project Authorization
PATU	Pan-African Telecommunication Union
PC/WATTC-88	Preparatory Committee for the 1988 World Administrative Telegraph and Telephone Conference
pdISP	Proposed Draft ISP
POSI	Promoting Conference for Open Systems Interconnection
PPSC-IT	Public Procurement Sub-Committee in Information Technology
PROVE	Provision of Verification
PSDN	Packet-Switched Data Network
PTC	Pacific Telecommunications Council
PTT	Posts, Telegraphs, and Telephones
R&D	Research and Development
R/C	Radio Control
RACE	Research and Development in Advanced Communication Technologies for Europe
RARE	Réseaux Associés pour la Recherche Européenne
RASCOM	Regional African Satellite Communication
REMCO	ISO Council Committee on Reference Materials
RF	Radio Frequency
RFI	Radio-Frequency Interference
RPOA	Recognized Private Operating Agency
RSVP	RACE Strategy for Verification and Plan
RW	Regional Workshop
RWS-CC	Regional Workshop Coordinating Committee
SAA	Standards Association of Australia
SAC	Standards Advisory Committee
SAE	Society of Automotive Engineers
SAG	Special Autonomous Group (*see also* GAS)
SANZ	Standards Association of New Zealand
SC	Subcommittee
SCC	Standards Coordinating Committee

SCC	Standards Coordinating Committee
SCC	Standards Council of Canada
SDR	Special Drawing Right
SG	Study Group
SG-FS	Special Group on Functional Standardization
SIG	Special Interest Group
SIM	Standards Institution of Malaysia
SIRIM	Standards and Industrial Research Institute of Malaysia
SISIR	Singapore Institute of Standards and Industrial Research
SITA	International Society for Aeronautical Telecommunications
SME	Society of Manufacturing Engineers
SNA	Systems Network Architecture
SNV	Association Suisse de Normalisation
SOG-ITS	Senior Officials Group for Information Technology Standardization
SOG-T	Senior Officials Group for Telecommunication
SPAG	Standards Promotion and Application Group
SPMC	Standards Program Management Committee
SRC	Strategic Review Committee
SS	Systems Support
STACO	ISO Council Committee on Standardization Principles
STAR	Special Telecommunications Action for Regional Development
SWG-P	Special Working Group on Procedures
SWG-RA	Special Working Group on Registration Authorities
SWG-SP	Special Working Group on Strategic Planning
SWIFT	Society for Worldwide Interbank Financial Telecommunications
TA	Technical Assembly
TAG	Technical Advisory Group
TBB	Transnational Broadband Backbone
TC	Technical Committee
TCD	Technical Cooperation Department
TCP/IP	Transmission Control Protocol–Internet Protocol
TEDIS	Trade Electronic Data Interchange Systems
TELDEV	Telecommunications Development in Six Arab States
TELECOM	Telecommunication Exhibition
TIA	Telecommunication Industry Association
TISI	Thai Industrial Standards Institute
TLFF	Technical Level Feeder's Forum
TOP	Technical and Office Protocol
TTC	Telecommunications Technology Council
TUANZ	Telecom Users Association of New Zealand

UA	User Agent
UAPT	African Telecommunication Union
UIT	Union Internationale des Télécommunications (see ITU)
UL	Underwriters Laboratories, Inc.
UNDAP	United Nations Development and Assistance Program
UNDP	United Nations Development Program
US CCITT	United States Organization for the International Telegraph and Telephone Consultative Committee
USTSA	US Telecommunication Suppliers Association
VAN	Value-Added Network
VT	Virtual Terminal
WAN	Wide Area Network
WARC	World Administrative Radio Conference
WARC HFBC-87	Second Session of the World Administrative Radio Conference for the Planning of the High Frequency (HF) Bands Allocated to the Broadcasting Service, Geneva, 1987 (*see also* HFBC-87)
WARC MOB-83	World Administrative Radio Conference for the Mobile Services, Geneva, 1983 (*see also* MOB-83)
WARC ORB-85	First Session of the World Administrative Radio Conference on the Use of the Geostationary-Satellite Orbit and the Planning of the Space Services Utilizing It, Geneva, 1985 (*see also* ORB-85)
WARC ORB-88	Second Session of the World Administrative Radio Conference on the Use of the Geostationary-Satellite Orbit and the Planning of the Space Services Utilizing It, Geneva, 1988 (*see also* ORB-88)
WATTC	World Administrative Telegraph and Telephone Conference
WG	Working Group
WOS	Workshop on Open Systems

The Artech House Telecommunication Library

Vinton G. Cerf, *Series Editor*

A Bibliography of Telecommunications and Socio-Economic Development
by Heather E. Hudson

Advances in Computer Systems Security: 3 volume set, Rein Turn, ed.

Advances in Fiber Optics Communications, Henry F. Taylor, ed.

Broadband LAN Technology by Gary Y. Kim

Codes for Error Control and Synchronization by Djimitri Wiggert

Communication Satellites in the Geostationary Orbit by Donald M.
Jansky and Michel C. Jeruchim

Current Advances in LANs, MANs, and ISDN, B.G. Kim, ed.

Design and Prospects for the ISDN by G. DICENET

Digital Cellular Radio by George Calhoun

Digital Image Signal Processing by Friedrich Wahl

Digital Signal Processing by Murat Kunt

E-Mail by Stephen A. Caswell

Expert Systems Applications in Integrated Network Management, E.C.
Ericson, L.T. Ericson, and D. Minoli, eds.

Handbook of Satellite Telecommunications and Broadcasting, L. Ya.
Kantor, ed.

Innovations in Internetworking, Craig Partridge, ed.

Integrated Services Digital Networks by Anthony M. Rutkowski

International Telecommunications Management by Bruce R. Elbert

Introduction to Satellite Communication by Bruce R. Elbert

International Telecommunication Standards Organizations by Andrew
Macpherson

Introduction to Telecommunication Electronics by A.Michael Noll

Introduction to Telephones and Telephone Systems by A. Michael Noll

Jitter in Digital Transmission Systems by Patrick R. Trischitta and Eve L.
Varma

LANs to WANs: Network Management in the 1990s by Nathan J. Muller
and Robert P. Davidson

Long Distance Services: A Buyer's Guide by Daniel D. Briere

Manager's Guide to CENTREX by John R. Abrahams

Mathematical Methods of Information Transmission by K. Arbenz and
J.C. Martin